ENVIRONMENTAL STRESS
AND
CELLULAR RESPONSE
IN ARTHROPODS

T0187630

André Korsloot
Cornelis A.M. van Gestel
Nico M. van Straalen

CRC Press
Taylor & Francis Group
Boca Raton London New York

CRC Press is an imprint of the
Taylor & Francis Group, an **informa** business

CRC Press
Taylor & Francis Group
6000 Broken Sound Parkway NW, Suite 300
Boca Raton, FL 33487-2742

© 2004 by Taylor & Francis Group, LLC
CRC Press is an imprint of Taylor & Francis Group, an Informa business

First issued in paperback 2019

No claim to original U.S. Government works

ISBN-13: 978-0-367-45438-8 (pbk)
ISBN-13: 978-0-415-32886-9 (hbk)

Visit the Taylor & Francis Web site at
http://www.taylorandfrancis.com

and the CRC Press Web site at
http://www.crcpress.com

Library of Congress Card Number 2003069758

Library of Congress Cataloging-in-Publication Data

Korsloot, André.
  Environmental stress and cellular response in arthropods / André Korsloot, Cornelis A.M. van Gestel, and Nico M. van Straalen.
    p. cm.
  Includes bibliographical references and index.
  ISBN 0-415-32886-1
  1. Stress (Physiology) 2. Pathology, Molecular. 3. Soil pollution. 4. Arthropoda. I. Gestel, Cornelis A. M. van. II. Straalen, N. M. van. III. Title.

  QP82.2.S8K676 2004
  571.9'5--dc   222003069758

Preface

We analyzed the stress imposed by soil pollution from the perspective of animal ecology concentrating on terrestrial invertebrates. When studying the subject of stress in animals, it became clearer and clearer how broad this subject is and what a great impact it has on animal life. We further noticed that up to now the information on this subject has been rather fragmentary and lacks an up-to-date overview and analysis. This book aims to fill this gap. We realized, however, that in order to make a solid contribution to this field of science, we had to establish boundaries for the types of stresses and the animal species. We therefore concentrated on the interaction between arthropods and their environment and the relationship with environmental stress. Consequently, we had to exclude certain aspects from this review, such as pathological and other endogenous factors causing stress.

Although environmental stress is primarily related to the fitness and subsequent performance of the whole organism, the underlying biochemical reactions and physiological effects occur at the cellular level. To present the differences in effects from the various environmental stressors, this book provides a thorough analysis of these underlying mechanisms and how various cellular defense systems respond to different stressors in order to restore cellular homeostasis. When we analyzed these effects and responses, it became apparent that the individual systems are interdependent and cooperate in their responses, leading to an integrated reaction. This topic therefore merits special attention and is extensively reviewed in this book. Sophisticated defense systems notwithstanding, a return to cellular homeostasis is not always possible, and therefore issues such as cell death, tolerance, and aging are discussed in the second part of the book. The latter processes exhibit their effects at the organismal level. In this way, a connection between physiology and ecology is established.

This book describes environmental stress in arthropods and tries to analyze this process in all its aspects, from biochemical mechanisms to effects on the whole organism. We hope it will attract the interest of a variety of scientists, such as biochemists, physiologists, cell biologists, and ecologists.

We thank Marianne H. Donker, Paul. J. Hensbergen, Jan E. Kammenga, Heinz R. Köhler, Jos A.C. Verkleij, and Saskia M. van der Vies for their comments on an earlier version of the manuscript. We also thank Els A. Korsloot-van Gemerden for preparing the manuscript, Paul Robert Korsloot and Nico Schaefers for the artwork, and Désirée Hoonhout for general assistance.

André Korsloot
Cornelis A.M. van Gestel
Nico M. van Straalen

The authors

André Korsloot, Ph.D, a retired technical engineer, completed his study of advanced technology in 1960 and received his degree in engineering (B.Sc.). He worked for 15 years as a design and process engineer in engineering offices, and for 17 years in process technical and managerial functions in oil refining.

After his retirement, Dr. Korsloot studied Environmental Sciences and received his M.Sc. from The Dutch Open University, Heerlen, the Netherlands, in 1997. In 2002 he obtained his Ph.D. from the Vrije Universiteit in Amsterdam. At present, he is affiliated with the Department of Animal Ecology, Institute of Ecological Science, of the Vrije Universiteit.

Cornelis A.M. van Gestel, Ph.D., is Associate Professor of Ecotoxicology at the Vrije Universiteit in Amsterdam. Dr. van Gestel received his M.Sc. (cum laude) in Environmental Sciences from the Agricultural University, Wageningen, the Netherlands, in 1981. From 1981 to 1986, Dr. van Gestel worked as a scientific advisor on the ecotoxicological risk assessment of pesticides at the National Institute of Public Health and Environmental Protection in Bilthoven, the Netherlands. In 1986, he became head of the Department of Soil Ecotoxicology of the same institute. He obtained his Ph.D. from Utrecht University in 1991. Since 1992, he has been affiliated with the Department of Animal Ecology, Institute of Ecological Science, of the Vrije Universiteit in Amsterdam.

Dr. van Gestel's major research interests are the toxicity and bioaccumulation of heavy metals and polycyclic aromatic hydrocarbons in soil invertebrates (Collembola, earthworms, isopods) in relation to routes of uptake and bioavailability in soil, and the effects at different levels of biological organization (biochemical, individual, population, community). He is also interested in issues of risk assessment and the use of bioassays to determine the actual ecological risk of contaminated land.

Dr. van Gestel is a member of the Society of Environmental Toxicology and Chemistry (SETAC) and the Netherlands Society of Toxicology. He is registered as a toxicologist by the Netherlands Society of Toxicology and the European Society of Toxicology (EUROTOX).

Dr. van Gestel is author or co-author of more than 100 papers and book chapters. He has been a member of several national committees of the

Netherlands Health Council advising on ecotoxicological issues. He serves on the editorial boards of the journals *Ecotoxicology, Environmental Pollution,* and *Environmental Toxicology and Chemistry.*

Nico M. van Straalen, Ph.D., is Professor of Animal Ecology at the Vrije Universiteit in Amsterdam and Director of the SENSE Research School for Socio-Economic and Natural Sciences of the Environment, a joint venture of eight Dutch universities.

Dr. van Straalen received his M.Sc. in 1979 and his Ph.D. in 1983 from the Vrije Universiteit in Amsterdam. He was a Lecturer in Ecology (1982–1988) and then Associate Professor of Ecotoxicology at the Department of Animal Ecology, Institute of Ecological Science of the Vrije Universiteit in Amsterdam, before he obtained a full professorship in Animal Ecology in the same department in 1992.

Dr. van Straalen developed a research program on ecotoxicology of soil invertebrates, focusing on toxicity, kinetics, mode of action, and adaptation of heavy metals and PAH for soil invertebrates (springtails, isopods, oribatids). Currently, his major reseach interest is the evolution of metal tolerance, including the application of molecular-genetic tools to unravel tolerance mechanisms in soil arthropods. He is also interested in issues of ecotoxicological risk assessment.

Dr. van Straalen is a member of Society of Environmental Toxicology and Chemistry (SETAC), the Netherlands Society of Toxicology, the Netherlands–Flemish Society of Ecology, the Netherlands Society of Entomology, and the British Ecological Society. He was a member of the SETAC-Europe Council (1988–1992) and chair of the SETAC-Europe Education Committee (1990–1996). He chaired the scientific committee for SETAC-Europe annual meetings in Sheffield (1991) and Amsterdam (1997), and was an invited lecturer for several international advanced courses on Ecotoxicology and Soil Ecology. Dr. van Straalen is registered as a toxicologist by the Netherlands Society of Toxicology and the European Society of Toxicology (EUROTOX). He received the ABC Laboratories/SETAC-Europe Environmental Education Award in 1998.

Dr. van Straalen is author or co-author of more than 150 papers, books, and book chapters. He is a member of the editorial boards of the journals *Pedobiologia* and *Applied Soil Ecology.* Dr. van Straalen has served on various national committees of the Netherlands Health Council, the Netherlands Integrated Soil Research Programme, and the Advisory Board of the Dutch Research Programme on *in situ* Sanitation of Soils (NOBIS).

Contents

chapter 1

Introduction

1.1 What is stress?

In the environment, any organism is confronted with stress, for instance, stress caused by fluctuating or changing environmental conditions and disturbances by man or by pollution. Such stress may be harmful to biological life, exhibiting adverse effects at different levels of organization, such as populations, individual organisms, tissues, and cells. These effects are demonstrated in life-history traits of an organism, for example, development, growth, aging, longevity, survival, and reproduction. Biologists borrowed the term *stress* from physics, where it indicates force per unit area, so in this case stress is caused by an external agent acting on a system. In biology, however, stress became synonymous with the internal consequences of an external factor. There seems to be agreement now that a distinction should be made between *stressor* (an external factor), *stress* (an internal state brought about by a stressor), and the *stress response* (a cascade of internal changes triggered by stress). Although the concept of stress can be defined at various levels of biological integration, varying from cells, individuals, populations, and ecosystems (Parker et al., 1999), stress is most commonly studied in the context of individual organisms (Maltby, 1999), and the responses on the level of cells and biochemical pathways (as covered in this book).

The concept of stress is not absolute. What is an extremely stressful condition for one organism (e.g., salt water for a freshwater fish) is quite normal for another organism (salt water for marine fish). So stress must be defined in terms of a species' ecological niche. A definition of stress that incorporates this idea runs as follows:

> Stress is a condition evoked in an organism by one or more environmental factors that bring the organism near to or over the edges of its ecological niche.

In addition, stress has the following properties:

1. It is usually transient.
2. It involves a syndrome of specific physiological responses.
3. It is accompanied by the induction of mechanisms that counteract its consequences.

This niche-based definition of stress is illustrated schematically in Figure 1.1 (van Straalen, 2003). Stress arises when some environmental factor changes such that an organism finds itself outside its ecological niche. Outside the niche, the organism (by definition) cannot grow and reproduce, but it may survive temporarily. There are two options to relieve the stress: (1) moving back to the niche by using behavioral mechanisms or by suppressing the stressor and (2) changing the boundaries of the niche by genetic adaptation. The first option must be accompanied by temporary physiological adaptation, which allows survival until the stressor has gone. Calow recognized these two responses (1989) as the *proximate* and *ultimate* responses to stress. The stress responses described in this book are all proximate defense mechanisms, whose final aim is to cope with the stressor by minimizing or repairing damage.

The concept of niche must be specified further, because it is often not a property of a species as a whole but may also vary between populations of a single species. In that case, a species with a wide ecological amplitude (a *euryoecious* species) may consist of several local populations, each with a narrow amplitude (*stenoecious* populations). Consequently, what is experienced as stress for one population may be normal for another population of the same species. Furthermore, even the same population may harbor different genotypes that have different ways of dealing with stress. In a polymorphic population, natural selection may favor one particular way of dealing with stress over others and thus change the genetic make-up of the population. Ultimately, a population may become fixed for one particular stress response, and this is designated as *genetic adaptation*. The study of such adaptations is an important topic in evolutionary ecology and physiology (Hoffmann and Parsons, 1991; Parsons, 1997; Feder et al., 2000). Of special interest are the secondary consequences of adaptation, which can often be viewed as trade-offs, because they represent metabolic or other costs that the organism has to expend in order to permanently cope with stress. This book focuses on proximate mechanisms and only lightly touches upon the phenomenon of genetic adaptation.

Whatever the level of exhibition of damage or whatever the trait affected, the basic attack and subsequent physiological changes take place in the organism's cells. At this level of attack, stress may be considered a disturbance of cellular homeostasis, which may lead to cell death, impaired functioning, or return to homeostasis, depending on the type and severity of the stress and the cellular stress response. The latter depends on the activity of cellular defense systems, which may vary by cell type. The study of stress responses therefore may reveal important knowledge about how cellular processes are regulated. In addition, there are many interactions between

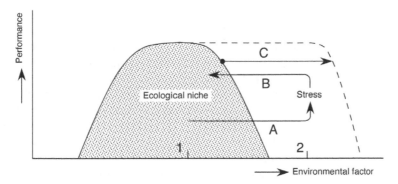

Figure 1.1 Graph showing a definition of stress based upon the ecological niche of a species. (A) Stress arises when an environmental factor increases from 1 to 2 such that the species finds itself out of its ecological niche; (B) various stress response reactions allow temporary survival under stress, but this is followed by a return to the niche; (C) alternatively, the borders of the niche are extended by adaptation, and what was formerly stress is not stress anymore. (Reproduced from van Straalen, N.M., *Environ. Sci. Technol.* 37: 324A–330A, 2003. With permission from the American Chemical Society.)

the stress response systems directed at different types of stress. The effect of these interactions is that the cell acts as an integrated unit to maintain homeostasis.

1.2   *Environmental stressors*

There are two major sources of cellular stress in an organism. The first source is endogenous factors, mainly of a biological origin, such as pathogens. The second source is environmental factors, which may have a physical character, such as heat, cold, drought, and osmotic conditions, or may be of a chemical nature, such as oxidants, heavy metals, and organic chemicals. Chemical factors may cause cellular stress either by transferring a signal to the cell or by penetrating the cell itself. Both sources of stress may influence cellular homeostasis in a way that impairs proper functioning of the cell. This book will concentrate on environmental factors and their effects and only refer to endogenous factors and their effects if necesssary for a good understanding of the mode of action of the environmental factors (see Section 1.5).

Physical and chemical stressors may cause so-called environmental stress, which includes more specific stresses, such as oxidative stress, metal stress, and genotoxic stress, in relation to the type of stressor or the specific effect exerted. Indeed, the various types of stress stem from different types of stressors that affect cellular integrity and functioning in different ways. These stressors can be subdivided into the following six groups, depending on the mode of action and the type of effect exerted:

1. Stressors inhibiting or affecting proper protein synthesis, such as amino acid analogs, cycloheximide, and ethanol
2. Stressors causing cellular stress of a more general nature, such as heat, which leads to the induction of so-called stress proteins
3. Special stressors affecting specific cellular functions but also inducing the synthesis of stress proteins (e.g., ionophores affecting cellular ion homeostasis and a treatment leading to cellular glucose deprivation)
4. Stressors causing oxidative stress, including oxidants such as hydrogen peroxide, redox cyclers such as paraquat, and quinonoid compounds such as benzoquinone
5. Stressors causing metal stress, such as cadmium and arsenite
6. Stressors causing genotoxic stress, such as hydrogen peroxide, arsenite, and irradiation by ultraviolet light, and organic chemicals or their metabolites that have a high DNA binding affinity

Since groups 1, 2, and 3 eventually exhibit similar effects by modifying cellular proteins and enzymes and by inducing stress proteins, they are jointly dealt with as stressors causing "proteotoxicity." This list of stressor types is not complete or exhaustive. It is an attempt to structure the different effects produced or induced by a variety of stressors. The differentiation among the groups cannot be distinct, since some stressors fit in more than one group and their effects may overlap. The groups of stressors in relation to the stress exerted and subsequent stress effects are shown in Figure 1.2. The stress effects will be discussed in the next section.

1.3   Stress effects disturbing cellular homeostasis

Depending on the type of stress, its severity, and the duration of the stress event, a cell may be confronted with increasing levels of effects and damage from negligible to detrimental to its viability. In the literature, these levels of damage are divided into three stages:

1. Damage that can be repaired by the cell's defense systems. Cellular homeostasis can be restored.
2. An intermediate level of damage. The cell is only able to partly repair damage and restore homeostasis but remains viable.
3. A high level of damage that cannot be repaired. Essential processes are inhibited and structures lose their integrity. The cell enters the process of apoptosis or dies by necrosis.

Apart from these three levels of damage in relation to cell status, the process of damage in relation to time and progress is described. On this basis, three levels of effects are distinguished:

1. Primary effects occurring at the molecular level, which may include synthesis of aberrant polypeptides, oxidation and denaturation of

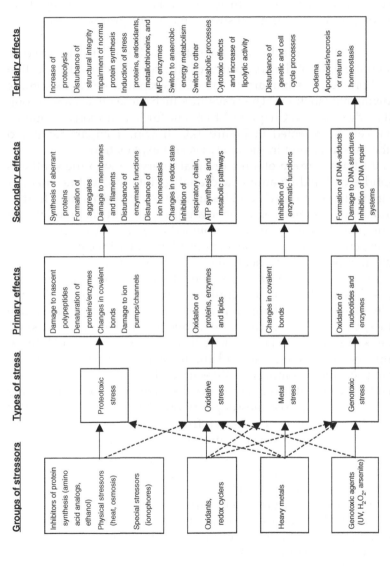

Figure 1.2 Summary of responsible groups of stressors, types of stress, and their effects on cellular integrity and functioning. Solid lines: relationships described in Chapter 1; dashed lines: relationships mainly described in Chapters 2 through 6.

structural proteins, changes in covalent bonds of enzymes, and oxidation of lipids and nucleotides

2. Secondary effects, which refer to damage to structures, such as membranes, DNA, and filaments, and effects on processes, such as respiration and energy production

3. Tertiary effects, which refer to the consequences of damage or effects on structures and processes, including collapse or loss of integrity of structures, inhibition of downstream metabolic processes, etc.

The effects of stress on cellular homeostasis will now be discussed in more detail on the basis of the six groups of stressors mentioned in Section 1.2. Figure 1.2 provides a summary of the discussion and serves as a guide for the following explanations. The ways in which these effects are accomplished will be discussed in Chapters 2 to 6. For further support, an outline of an animal cell is presented in Figure 1.3, showing compartments and organelles that are discussed in the analyses and explanations.

Figure 1.2 shows a group of stressors that intervene with the synthesis of (poly)peptides by building amino acid analogs into the peptide chain or by impeding synthesis of (poly)peptides in another way. This may lead to the production and accumulation of aberrant proteins and enzymes, which have to be degraded by an activated proteolytic system (Ananthan et al., 1986), at the expense of energy in the form of adenosine triphosphate (ATP).

Physical stressors, such as heat, show a broader field of action and damage (Figure 1.2). Heat shock causes denaturation of existing proteins and enzymes, leading to their accumulation and aggregation. This may trigger a series of events. It will impair proper functioning of structures, such as membranes, by increased permeability, causing (ion) leakage. At sufficiently severe stress, intracellular filament structures will be affected. Intranuclear filaments and filaments around the nucleus in particular may collapse, forming insoluble complexes (Bensaude et al., 1990). This may be accompanied by the formation of aggregates dispersed in the nucleus. These aggregates contain unprocessed messenger ribonucleic acid (mRNA). In this way, many mRNAs may be protected from degradation (Parsell and Lindquist, 1994). These events will also affect the functioning of the nucleolus by aggregation of preribosomes and ribonucleoprotein (RNP) particles (Morimoto et al., 1990b; Welch, 1990).

Inhibition of enzymes by conformational changes results in many kinds of effects within the cell, ranging from improper functioning of receptors, ion channels, and ATPase pumps in membranes to impairment of enzymatic functions in respiratory and metabolic processes (see Figure 1.2).

Accumulation of denatured proteins triggers the induction of stress proteins (Lindquist, 1986). This will be further discussed in Chapter 3, but what is of importance here are the side effects of this induction on the synthesis of "normal" proteins. Indeed, at induction of stress proteins, synthesis of normal proteins may be impaired at both the transcriptional and the translational level. In heat-shocked cells, gene expression of normal

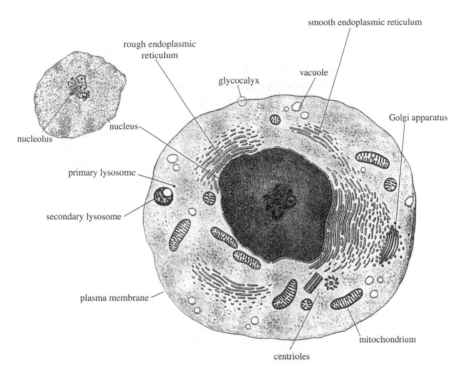

Figure 1.3 A typical animal cell. Not every animal cell contains all the organelles shown, nor are all the substructures that may occur in an animal cell represented, but the organelles mentioned in this book are shown. (Adapted from Keeton, W.T. and Gould, J.L., in *Biological Science*, 5th ed., Norton, New York, 1993.)

proteins is decreased by des-ubiquitination of histones and subsequent condensation of chromatin (Schlesinger, 1990); splicing of precursor mRNAs is interrupted in stressed cells, probably owing to conformational changes in spliceosomes (Yost et al., 1990); at the translational level, cap binding of the message leader of RNA, necessary for an efficient translation by ribosomes, is interrupted, probably by dephosphorylation of the cap-binding factor (Bensaude et al., 1996). The net effect is that, compared to normal proteins, stress proteins are preferentially synthesized during stress.

Returning to protein damage and inhibition of enzymes, the most dramatic secondary effects are probably disturbances in the respiratory chain and uncoupling of oxidative phosphorylation in the mitochondrium, because of destabilization of the integrity of its inner membrane (see Figure 1.2). The uncoupling effect may result in a drop in ATP level (Welch, 1990). This could lead to nicotinamide adenine dinucleotide hydride (NADH), the short-term buffer of energy in the cell, being unable to unload its electrons. As a consequence, aerobic energy metabolism ceases, followed by a shift from aerobic to anaerobic energy production (Welch, 1992). With the high energy demand under stressful conditions, this switch, accompanied by a lower ATP yield, is enforced and can be considered an escape from further detrimental effects of stress. The reduction in ATP level by severe heat shock and other stressors, and the increase in adenosine monophosphate (AMP) concentration, activates the AMP kinase cascade, which in turn inactivates synthesis of glycogen and fatty acids. Thus, a drop in ATP level may trigger an emergency shut-off of energy-consuming metabolic pathways (Bensaude et al., 1996). This is considered a tertiary effect (see Figure 1.2).

Another physical stress may be caused by cold during winter and in spring and autumn during short-term chilling periods. If protection against this damage is insufficient, membrane damage is the main consequence. Insufficient protection against chilling may lead to a cold shock causing a phase transition in membrane lipids and subsequent leakage. Insufficient protection against extracellular freezing may lead to membrane damage by dehydration and by redistribution of membrane components upon thawing, when the cell is in a hypertonic state.

Special chemicals, such as ionophores, also have a great impact on cellular homeostasis. Although these chemicals affect special targets, they have been included in the group of stressors causing proteotoxic stress because their effects also induce stress proteins. Ionophores may affect channels and ATPase pumps (Na^+/K^+ and Ca^{2+}) in the plasma membrane, as well as Ca^{2+}-ATPase pumps in membranes of the nucleus and the endoplasmic reticulum (ER). Disturbance of calcium homeostasis has several important effects, putting a cell's viability at stake. Mitochondrial Ca^{2+} release may deplete ATP levels, since Ca^{2+} is recycled over the inner membrane of the mitochondrium at the expense of ATP. Ca^{2+} release could be a starting point in the process leading to apoptosis (Fuchs et al., 1997; Goossens et al., 1995; Richter and Schweizer, 1997). According to Li and Lee (1991), depletion of sequestered Ca^{2+} in the ER blocks protein glycosylation, resulting in the induction of stress proteins. Release of Ca^{2+} from the organelles leads to increased Ca^{2+} concentration in the cytosol, affecting second messenger systems and, indirectly, gene expression.

Most oxidants, including redox cyclers and xenobiotics, such as quinonoid compounds, exert their oxidizing effect via the generation of a superoxide anion radical. If this radical cannot be removed quickly enough, it may oxidize sulfhydryl groups of proteins, enzymes, and antioxidants, such as glutathione (GSH). In this way, proteins may undergo conformational changes, enzyme activity may be inhibited, and the redox state of the cell may be shifted toward an oxidizing environment. Together, stressors of this type may cause oxidative stress (Wolin and Mohazzab-H, 1997; Yu, 1994). The effects are summarized in Figure 1.2.

Further processing of the superoxide anion radical and other intermediate compounds may yield, in a side reaction, the hydroxyl radical. If this radical is not effectively scavenged, lipid peroxidation may occur, leading to membrane damage and cytotoxic effects and finally to cell death (Goossens et al., 1995; Jacquier-Sarlin and Polla, 1996). At high levels of hydroperoxides in the membrane, lipolytic enzymes are activated to remove the oxidized lipids (Rice-Evans, 1994). Too high concentrations of lipid hydroperoxides affect the proper functioning of membranes. In the plasma membrane, the Na^+/K^+-ATPase pump is strongly affected by lipid hydroperoxides. This implies that severe oxidant stress is usually associated with cellular edema (see Figure 1.2). Voltage-dependent Ca^{2+} channels will be activated, leading to a sustained increase of intracellular Ca^{2+} (Chaudère, 1994). Lipid hydroperoxides also hamper the functioning of mitochondria by altering membrane permeability for various cations, resulting in an inhibition of oxidative phosphorylation (Chaudère, 1994). This has downstream effects on metabolic processes. Together, the effects resulting from severe oxidative stress may trigger the process leading to cell death (see Figure 1.2).

Metal stress is caused by heavy metals, such as cadmium, by changing covalent bonds in proteins, and by replacing zinc in the active sites of zinc-containing enzymes. In this way, metal stress may show the same effects as those described above when discussing damage to proteins and enzymes. Metal stress also exhibits genotoxic effects by inhibiting enzymes involved in DNA repair (Beyersmann and Hechtenberg, 1997; Ramos-Morales and Rodriguez-Arnaiz, 1995). Similar to heavy metals, other genotoxic agents, such as hydrogen peroxide, may cause genotoxic stress. The pathway and the sites of attack may be slightly different, however. In the case of hydrogen peroxide, hydroxyl radicals are probably generated, which may oxidize nucleotides and subsequently cause strand breaks. Strand breaks caused by heavy metals may be an indirect effect of oxidative stress participation (Beyersmann and Hechtenberg, 1997; Kawanishi, 1995). Many organic chemicals may also cause genotoxic effects, either directly or after activation by the P450 biotransformation system. Binding to DNA of these chemicals or their biotransformation products results in the formation of DNA adducts. The same (and some other) chemicals may also disturb other cellular processes by binding to proteins (protein adducts) (see Figure 1.2).

In summary, the type of stress is determined by the type of stressor. Because several effects are not specifically related to one type of stress, different stressors can evoke similar effects. The final effects depend on the severity of the stress and the cellular response to it. Three phases can be considered in the process from attack to final event: first, the biochemical reactions; second, the effects on intracellular processes; and third, the overall effect on cellular integrity and homeostasis. If the stress is so severe that stress defense systems cannot abolish the stress effects, the cell may enter a process leading to apoptosis or die by necrosis.

1.4   *Stress defense systems and their response to stress*

To cope with these disturbances and restore homeostasis, the cell possesses various defense systems with more or less specific tasks depending on the type of stressor. The systems discussed in this book are:

- The stress proteins
- The oxidative stress response system

- The metallothioneins
- The mixed function oxygenase system
- The basal signal transduction systems

Under steady state conditions, these systems play roles in cellular growth and proliferation and contribute to cellular homeostasis. Stress proteins are involved in folding, assembling, and intracellular transport of nascent proteins. The components of the oxidative stress response system decompose and scavenge reactive intermediates produced in respiratory and metabolic processes. Metallothioneins contribute to a cellular homeostasis of essential metals, whereas the mixed function oxygenase system is involved in the biotransformation and conjugation of xenobiotics, hormones, and metabolites.

Under stressful conditions, these systems fulfill extra tasks that are in line with the tasks performed under normal conditions. Stress proteins concentrate on protecting essential proteins to maintain cellular integrity. Components of the oxidative stress response system scavenge oxidants, maintain cellular redox states and initiate repair of oxidative damage. Metallothioneins participate in detoxifying heavy metals and in scavenging free radicals. Certain enzymes of the mixed function oxygenase system are activated if xenobiotic, hydrophobic chemicals penetrate the cell and cause damage. Stress signals are transduced via basal signal transduction systems to the transcriptional machinery in the nucleus of the cell, and these signal transduction systems therefore form an integral part of the cellular stress defense. In this way, the cell uses a complete package of measures to respond to environmental stress.

1.5 Scope and purpose of this book

This book describes how environmental stressors cause cellular stress and how the cellular stress defense systems respond. Apart from reviewing the functioning of these systems, emphasis is placed on the integrated response of these systems. Substantial support is provided for the hypothesis that _the individual stress defense systems are interrelated and cooperate in their response as an integrated cellular stress defense system._

After reviews of the individual systems in Chapters 2 through 6, support for this hypothesis will be provided in Chapter 9, in which the interrelationships between these systems and their integrated response are analyzed.

Chapters 7 and 8 discuss the physiological consequences of stress effects on tolerance, survival, cell death, and aging. By analyzing all these aspects in detail at the molecular, biochemical, and physiological level of the cell and by determining the consequences at the organismal level, this book will provide a comprehensive overview of the subject of stress and a thorough insight into the relationship of stress to the organism and its environment.

Our treatment of the material is as general as possible, but we have derived most of our inspiration and examples from the research on

Drosophila. This insect has become an important model for stress response studies, a situation that was reinforced by the complete sequencing of its genome (Adams et al., 2000). Most of the knowledge derived from *Drosophila* studies will also be valid for other organisms, and we illustrate this by occasionally referring to studies of yeast, nematodes, and vertebrates.

chapter 2

Basal signal transduction systems involved in stress response

2.1 Introduction

Basal signal transduction systems are usually involved in transducing extra-cellular signals into the intracellular environment and conveying the message to the site where the information is translated by converting the signal into an action. In the case of signals related to cell growth, differentiation, and proliferation, this often results in a transcriptional activation of genes related to these processes. Receptors in the plasma membrane, second messengers, such as cyclic adenosine monophosphate (cAMP) and Ca^{2+} ions, and protein kinase systems, such as protein kinase C (PKC) and mitogen-activated protein kinase (MAPK), belong to these signal transduction systems. The last step in signal transduction is often a phosphorylation of transcription factors. A discussion of these targets will be included in this chapter.

Components of the major stress defense systems are constitutively present to maintain cellular homeostasis and to contribute to cellular processes, such as growth, proliferation, and apoptosis. In these processes, the defense systems' actions are combined with those emanating from signals from, for example, growth factors and hormones via the signal transduction systems. For this reason, these basal signal transduction systems are also activated under stressful conditions, thus contributing to the stress response. They will be reviewed in this chapter in relation to their cooperation with stress response systems.

2.2 Two major second messenger systems: cAMP and Ca^{2+}

cAMP and Ca^{2+} are called second messengers because they transduce extra-cellular signals to effectors in the intracellular environment. These second messengers are reviewed together because, under steady conditions, both

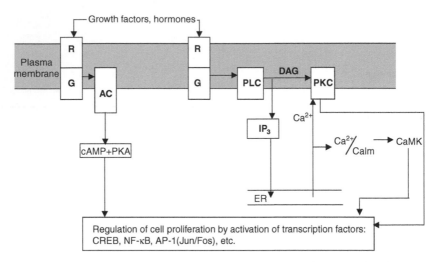

Figure 2.1 Simplified scheme for generating the second messengers cAMP and Ca^{2+}, as well as the activation of the kinases PKA and PKC by means of extracellular signals provided by growth factors and hormones. See text for further explanation. Abbreviations: R = receptor; G = G-protein; AC = adenylate cyclase; cAMP = cyclic adenosine monophosphate; PKA = protein kinase A; PLC = phospholipase C; DAG = diacylglycerol; IP_3 = inositol triphosphate; Calm = calmodulin; PKC = protein kinase C; ER = endoplasmic reticulum; CaMK = Ca^{2+} calmodulin-dependent protein kinase; CREB = cAMP-responsive element-binding protein; NF-κB = nuclear factor κB; AP-1 = activator protein-1.

systems are involved in the regulation of cell proliferation (Burdon, 1994). The systems will be described on the basis of Figure 2.1.

With regard to the "Ca^{2+} system" (belonging to the phosphoinositide system) (Calderwood and Stevenson, 1993), certain growth factors and hormones can interact with plasma membrane receptors (R), which activate phospholipase C (PLC) via a group of membrane-associated proteins (G-proteins). PLC brings about the formation of cytosolic inositol triphosphate (IP_3) and membrane-bound diacylglycerol (DAG). IP_3 causes the release of calcium ions from the endoplasmic reticulum (ER). DAG and Ca^{2+} subsequently stimulate the activation of the membrane-bound protein kinase C (PKC) (Burdon, 1994; Chin and Means, 2000; Krauss, 2001).

With regard to the "cAMP system," the extracellular regulators communicate a stimulating or inhibiting signal to adenylate cyclase (AC), which is responsible for an increase or decrease in cyclic adenosine monophosphate (cAMP) and the activation of the cAMP-dependent protein kinase A (PKA) (Burdon, 1994).

Apart from numerous side effects, both itogenic systems are involved in cell proliferation by activating transcription factors and related gene expression. Via PKA, cAMP specifically activates the transcription factor cAMP-responsive element-binding protein (CREB) (Bollen and Beullens, 2002), which binds cAMP-responsive elements (CREs) in the promoters of genes involved in DNA synthesis during the S-phase (Burdon, 1994; Hill and Treisman, 1995). Sheng et al. (1991) demonstrated, however, that CREB is also activated by phosphorylation by Ca^{2+}-calmodulin-dependent protein kinase (CaMK) of the phosphoinositide system. This was also reported by Krauss (2001) and Zaccolo et al. (2002). Thus, CREB functions by integrating Ca^{2+} and cAMP signals. However, continuous exposure of certain cells to cAMP can inhibit the cell division process. The effects of cAMP may therefore depend on the differentiation state of the cell (Burdon, 1994). This aspect is important in relation to the effects of environmental stressors on the cAMP system.

The activation of certain transcription factors by the Ca^{2+}/calmodulin system is mainly effected via PKC (Burdon, 1994). This important protein kinase is involved in the induction of c-Jun and c-Fos proteins, which form the dimeric transcription factor activator protein-1 (AP-1) (Flores and McCord 1997; Karin, 1995; Storz and Polla, 1996). AP-1 induces gene expression for cell proliferation but is also activated under stressful conditions, contributing in this way to the stress response (Hill and Treisman, 1995; Karin, 1995). PKC participates in a variety of other activities: It influences posttranscriptional control (Guyton et al., 1997), it effects the cAMP system by activating AC, and it has both activating and inhibiting effects on the MAPK cascade (Burgering and Bos, 1995). PKC also activates the transcription factor nuclear factor-κB (NF-κB) (Hill and Treisman, 1995). This factor activates genes expressing mRNA for proteins involved in inflammatory responses, redox control, and cell cycle. NF-κB is also activated by environmental stressors (Storz and Polla, 1996). Calcium ionophore releases Ca^{2+} from the ER. The transiently increased Ca^{2+} concentration in the cytoplasm mediates the activation of PKC and subsequent transcription of transcription factors. Other compounds, such as the phorbol ester phorbol 12-myristate 13-acetate (PMA), are able to activate PKC in a direct way (Jacquier-Sarlin et al., 1995). The phosphoinositide system is also involved in the induction of stress response in an indirect way by heat shock, ethanol, and arsenite. Calderwood and Stevenson (1993) demonstrated that heat shock stimulates PLC activity to a level comparable to activation by growth factors. Subsequently, DAG and IP_3 are activated, followed by release of Ca^{2+} from the ER and activation of protein kinases, such as PKC and CaMK. Calderwood and Stevenson (1993) suggest that PLC activation upon heat shock is caused by guanosine triphosphate (GTP) binding to G-proteins in permeabilized membranes. All these examples of activating second messenger systems and subsequently transcription factors show how these second messenger systems are involved in transducing stress signals, resulting in a stress response.

2.3   The MAPK cascade

The MAPK cascade plays an integral role in the transduction of signals from both stress and mitogen stimuli, such as growth factors and hormones, culminating in the phosphorylation of nuclear factors and the transcriptional activation of downstream genes. MAPK family members interplay in cellular responses, from proliferation to apoptosis (Guyton et al., 1997; Kholodenko, 2002). Figure 2.2 shows a simplified scheme of the MAPK cascade.

The MAPK cascade consists of multiple pathways, which affect each other through various combinations. In higher animals, there are pathways that predominantly transduce either stress or mitogenic signals promoting cell proliferation. The pathway transducing mainly mitogenic signals is considered the classical one in metazoans. The phylogenetic span of the stress-signaling pathway has not yet been fully resolved. The enzymes mentioned in this pathway are probably found in higher eukaryotes. Thus, the high osmolarity gene 1 product (HOG1) in *Saccharomyces cerevisiae* corresponds with RK/p38 in higher organisms (Rouse et al., 1994), such as in *Drosophila melanogaster* (Han et al., 1998; Martin-Blanco, 2000). Most of the other enzymes shown in Figure 2.2 are also found in *D. melanogaster*, at least in a homologous form. For instance, Botella et al. (2001) demonstrated that the c-Jun N-terminal kinase (JNK) in *Drosophila* (DJNK) plays an important role in stress responses that mirrors its mammalian counterpart. Some of the enzymes have been given different names, which are mentioned in the figure between brackets and in italics. These names refer to the gene coding for these enzymes.

The classical MAPK pathway is stimulated by growth factors, cytokines, tumor necrosis factor (TNF)-α, hormones, etc., through activation of receptor tyrosine kinases (RTKs). In this pathway, RTKs activate downstream Ras, or directly activate Raf enzymes, leading to amplification of the signal and activation of the effector, the extracellular signal-regulated kinase (ERK). Avruch et al. (1994) emphasize the central role of Ras and Raf in *Drosophila* in cellular differentiation and DNA synthesis. Perrimon (1994) described several RTK types in *Drosophila*.

The stress signal may be transduced along a special pathway but may also affect enzymes of the classical pathway. Oxidative stress in particular may lead to oxidation of receptors and critical sulfhydryl groups in phosphatases, such as MAPK phosphatase (MKP), indirectly increasing kinase activity. Cavigelli et al. (1996) reported an example of MAPK phosphatase inhibition. They noticed that arsenite stimulates AP-1 activity by inhibiting a JNK phosphatase. The JNK phosphatase has a dual specificity and also regulates the RK/p38 enzyme activity. Both JNK and RK/p38 stimulate AP-1 activity, leading to increased transcription of the c-Jun and c-Fos proto-oncogenes and possible carcinogenic effects. Activated PKC may also amplify the MAPK signal (Guyton et al., 1997; Holbrook et al., 1996; Wolin and Mohazzab-H, 1997).

Figure 2.2 only shows effects of interest from a stress-response point of view. Examples are activation of small heat shock proteins (discussed in Chapter 3) and transcription factors. Bensaude et al. (1996) reported that the small heat shock proteins are activated by the enzyme reactivating kinase (RK), which is equivalent to the p38 enzyme, mentioned above (see Figure

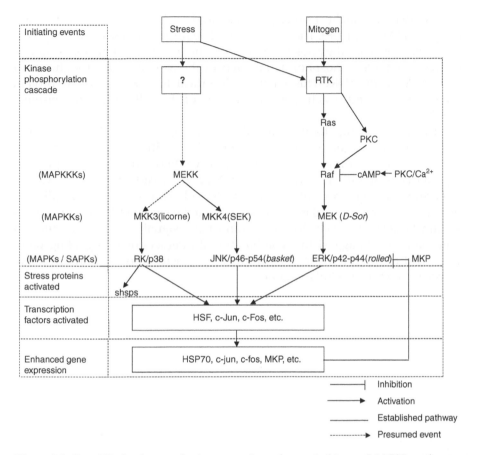

Figure 2.2 Simplified scheme of mitogen-activated protein kinase (MAPK) pathways for stress and mitogens. The two pathways are not independent and may influence each other. The Ras-ERK pathway is considered the classical one in metazoans. The phylogenetic span of the stress pathway is not yet fully resolved. The effects in relation to stress mainly implicate the activation of small heat shock proteins and transcription factors. Solid lines indicate established pathways, dashed lines represent presumed but less well-established events. Abbreviations: RTK = receptor tyrosine kinase; PKC = protein kinase C; cAMP = cyclic adenosine monophosphate; MEK = MAPK kinase (a MAPKK); MEKK = MEK kinase (a MAPKKK); ERK = extracellular signal-regulated kinase; p42–p44 are ERK isoforms; MKK 3/4 = MAPK kinase 3/4; SEK = SAPK kinase; SAPK = stress-activated protein kinase; RK = reactivating kinase; p38 = equivalent to RK; JNK = c-Jun N-terminal kinase; p46–p54 are JNK isoforms; HSF = heat-shock factor; HSP = gene for heat-shock protein; shsps = small heat-shock proteins; MKP = MAPK phosphatase gene/product; the words in italics refer to genes in *Drosophila melanogaster* that express proteins functionally corresponding to the ones mentioned in the scheme. (Adapted from Avruch, J. et al., *Trends Biochem. Sci.* 19: 279–283, 1994; Burgering, B.M.T. and Bos, J.L., *Trends Biochem. Sci.* 20: 18–22, 1995; Holbrook, N.J. et al., in *Stress-Inducible Cellular Response* Birkhäuser, Basel, Switzerland, 1996; Krauss, G., in *Biochemistry of Signal Transduction and Regulation,* 2nd ed., Wiley-VCH, Weinheim, Germany, 2001; Marshall, C.J., *Cell* 80: 179–185, 1995; Martin-Blanco, E., *BioEssays* 22: 637–645, 2000; Rouse, J. et al., *Cell* 78: 1027–1037, 1994.)

2.2). Activated small heat shock proteins mediate the cellular redox state and apoptosis (Arrigo, 1998). These aspects will be further reviewed in Chapter 8.

In Section 3.3.1, the phosphorylation of the transcription factor c-Jun will be described. One phosphorylation step in the activation is brought about by a MAPK, the JNK. This kinase is identical to stress-activated protein kinase (SAPK) and is active in the nucleus of the cell (Hill and Treisman, 1995).

In the following chapters, it will become apparent that many stimuli induce the same signaling pathways yet activate different genes or activate the same genes at different strengths. To solve these problems, complicated regulatory systems are required. The MAPK cascade is such a system. According to Avruch et al. (1994), the MAPK cascade is so complicated because an array of regulatory inputs have to be signaled to various downstream targets, at which diversification of the signal is essential. Hill and Treisman (1995) suggest that combinational interactions of signaling pathways and transcription factors introduce specificity into the transcriptional response to (extra) cellular signals.

chapter 3

The stress-protein system

3.1 Introduction

To cope with environmental stress, all cells (prokaryotic and eukaryotic) possess the ability to induce so-called stress proteins (SPs). The stress proteins differ between phyla with regard to type, molecular weight, and expression, but their DNA sequences show that they belong to one gene family and that they have been highly conserved throughout evolution. Nucleotide homologies between stress proteins in prokaryotes and eukaryotes are 40 to 50% and between stress proteins in eukaryotes are in the range of 70% (Plesofsky-Vig, 1996). For instance, the human hsp70 is 73% identical to its equivalent protein in *Drosophila* and 50% identical to its equivalent protein (DnaK) in *Escherichia coli* (Lindquist, 1986).

Among the various types of stress protein families described below, the sp70 family is the most important. In *Drosophila*, it consists of two types of stress proteins: the so-called heat-shock cognates (HSCs) and the heat-shock proteins (HSPs). The HSCs are constitutively present, whereas the HSPs are induced by stress. This does not mean that HSPs are absent in unstressed cells, but their concentrations are often very low. Correspondingly, the cognates are also present in stressed situations and assist the HSPs in defending the cell against abuse. Some or all of the HSCs may even be moderately induced. Stress proteins of other families may be both constitutively present and induced under stressful conditions.

Under normal, unstressed conditions, the constitutively present stress proteins participate in protein folding and assembling, metabolic processes, and cell growth and development. These stress proteins are therefore essential for cell viability (Elefant and Palter, 1999; Ellis and van der Vies, 1991; Welch, 1993). The HSPs are induced under stressful conditions to defend the cell against external stressors. It is suggested that during stress, stress proteins protect essential proteins against denaturation, assist in repairing damage, and are possibly involved in the disassembly and degradation of abnormally folded proteins (Sanders, 1990; Welch, 1992).

If the cell is exposed to stress, its response is often very fast: Some stress proteins are induced within a few minutes, building up high concentrations within an hour (Lindquist, 1986). At the same time, normal functions are compromised so that all efforts can be directed toward survival and restoration of homeostasis. Whether this will happen depends mainly on the severity of the stress and the duration of the stress event. Lindquist (1986) has raised the question of how most of the agents mentioned in the Introduction of this book can provoke the same reaction by inducing stress proteins. Is there a common mechanism? Do they trigger the reaction at the same point? It is now believed that, although the biochemical mechanisms may differ, many of the above-mentioned stressors are able to denaturate or modify proteins. It has been proven that denaturated or aberrant proteins are able to induce the stress response. Studies by Kelley and Schlesinger in 1978 showed that adding amino acid analogs to animal cells was sufficient to initiate a stress response. Addition of amino acid analogs results in abnormally folded proteins. Welch (1993) reported that it was Hightower, who suggested in 1980 that the accumulation of denaturated or abnormally folded proteins in a cell initiated a stress response. A few years later, Voellmy and Goldberg tested and confirmed Hightower's proposal. In a landmark study, they showed that inserting denaturated proteins into living cells was sufficient to induce a stress response (Welch, 1993).

3.2   Types, structures, and functions of stress proteins

This section presents an overview of stress proteins in *Drosophila melanogaster*. The stress proteins of this species will also be compared with stress proteins in human cells, because research has been conducted on cultured human cells, especially in relation to (viral) infections and diseases. A selection of the most important stress protein types is presented in Table 3.1 and will be discussed below.

3.2.1   The sp70 family

The sp70 family is the largest group of stress proteins and plays a central role in defending the cell against stress. As illustrated above, members of the sp70 family have been highly conserved throughout evolution. Compared to identical human stress proteins, there are regions of extraordinary conservation, predominantly in the N-terminal domain, and less-conserved regions in the C-terminal domain. Concerning the diverging regions in the amino acid sequence, an unanswered question concerns what changes are due to flexibility in the sequence and what changes reflect selected biological adaptations (Lindquist, 1986). In *Drosophila*, members of the sp70 family may differ considerably in homology depending on their location and function in the cell. Rubin et al. (1993) made a pairwise comparison of the members of the sp70 family of *D. melanogaster* on the basis of the aa sequences of these stress proteins. They demonstrated 72 to 82% homology

Table 3.1 Families and types of stress proteins found in *Drosophila melanogaster*

Family	Types	Chromosomal site (gene number)	Location		Human equivalent	Function
			unstressed	stressed		
Sp90	Hsp83	63BC	Cytoplasm	Cytoplasm	hsp90	Binds specific proteins (e.g., hormone receptors) and chaperones polypeptides to membranes
Sp70	Hsp70	87A7/87C1	Nucleus (cytoplasm)	Nucleus Nucleolus Cytoplasm	hsp72	Inducible component of the stress response
	Hsp68	95D	Nucleus; cytoplasm?	Nucleus; cytoplasm?	?	Inducible type; function unknown
	Hsc70	88 E (HSC4)	Cytoplasm	Cytoplasm; nucleus	hsp73 (hsc70)	Constitutive protein binding component in cytoplasm
	Hsc70b	70C (HSC1)	Cytoplasm	Cytoplasm (nucleus)	?	Constitutive protein-binding component involved in all stages of development
	Hsc71	50 E/5C (HSC5)	Mitochondrium	Mitochondrium	grp75	Constitutive protein-binding component in mitochondrium
	Hsc72	10 E (HSC3)	Endoplasmic reticulum	Endoplasmic reticulum	grp78 (BiP)	Constitutive protein-binding component in endoplasmic reticulum
	?	87D/(HSC2)	?	?	?	?
Small hsp's	Hsp22,23 hsp26,27	67BC	Cytoplasm	Nucleus (cytoplasm)	?	Participate in development and in stress response
Ubiquitin	Ubiquitin	?	Nucleus; nucleolus; cytoplasm	Cytoplasm (nucleus)	Ubiquitin	Tags protein for degradation

Note: SP = stress protein; HSP = heat-shock protein; HSC = heat-shock cognate; GRP = glucose regulated protein. Uppercase letters refer to genes, lowercase letters to proteins. Locations in parentheses mean not yet fully proven.

between cytoplasmic stress proteins, but only 49 to 60% between cytoplasmic stress proteins and stress proteins in the organelles. This pattern of differentiation indicates specialization and supports the observation that the cytoplasmic stress proteins cannot substitute for the stress proteins in the organelles (see Section 3.4 for further discussion).

All hsp70 proteins contain specific functional domains. A conserved region in the N-terminal part contains an adenosine triphosphate (ATP)–binding domain, whereas the less conserved C-terminal region contains a peptide-binding site that may bind other peptides (Plesofsky-Vig, 1996). In fact, ATP is enclosed by two ATP-binding domains forming a cleft (Morimoto and Milarski, 1990). All members of the sp70 family have similar ATPase fragments and tertiary structure (McKay et al., 1994).

The above-mentioned stress proteins of the sp70 family can be divided into four groups: the cytoplasmic cognates (hsc70 and hsc70b), the mitochondrial stress protein (hsc71), the major stress protein in the ER (hsc72), and the inducible cytoplasmic (nuclear) stress proteins (hsp70 and hsp68).

3.2.1.1　The cytoplasmic cognates: hsc70 and hsc70b

Hsc70 is the main cytoplasmic cognate and is heavily concentrated around the nucleus and abundantly present during development and normal growth. It is analogous to hsc70 (hsp73) in human cells (see Table 3.1). Under stressful conditions, hsc70 is moderately induced, but its rate of induction is much lower than its inducible equivalent, hsp70. Hsc70b is the second cytoplasmic cognate, but is 30-fold less abundant than hsc70 (Rubin et al., 1993). It may be present at any or all stages of development or it might even be the prominent cytoplasmic stress protein in a specific tissue (Rubin et al., 1993). According to Pauli and Tissières (1990), its gene (HSC1) has one intron, and it is therefore questionable whether this stress protein is inducible under severe stress conditions (see Section 3.3 for further discussion).

Under normal conditions, the main functions of cytoplasmic cognates are:

1. Facilitating synthesis and folding of polypeptides
2. Assisting in assembling of folded polypeptides
3. Chaperoning precursor proteins across intracellular membranes

Facilitating synthesis and folding of polypeptides — Cytoplasmic cognates bind to the polypeptide as soon as the polypeptide's N-terminal leaves a ribosome. In this way premature folding of the emerging polypeptide is prevented and completion of synthesis facilitated. Partial folding takes place after release from the ribosome to bring the nascent protein in a competent state for either further assembling processes or for membrane translocation (Beckmann et al., 1990; Nelson et al., 1992).

Assisting in assembling of folded polypeptides — The same stress proteins participate in the formation of secondary and tertiary structures of the nascent proteins and in the assembly of protein complexes. For that purpose, the cytoplasmic proteins that bind to cognates contain multiple binding sites with neutral residues in heptamer sequences. Hightower et al. (1994) propose a two-step interaction between hsc70 and extended polypeptides, in which ionic forces provide the initial attraction between a negatively charged cluster on an α-helix in or near the binding site of hsc70 and positively charged residues in target amino acid sequences of the polypeptide (ionic attraction is omnidirectional and is exerted over a greater distance than any other chemical bond in an aqueous environment). In the second step, the binding of the peptide triggers a conformational change in the binding domain of hsc70, bringing the neutral residues of the peptide into a hydrophobic environment. This induced fit is then based on shorter-range van der Waals attractive forces and hydrogen bonding with the peptide backbone (Hightower et al., 1994). This interaction is schematically shown in Figure 3.1.

Release and folding of the peptide is considered to be a stepwise process. A programmed release from hsc70s could be achieved by strategic placement of sequences on the peptide that differ in dissociation times (Rothman, 1989). Rothman considers it a catalytic process, although the general opinion is that stress proteins do not act as catalysts but rather stabilize the peptides in their folding and assembly process (Feder, 1996; Gething and Sambrook, 1992). According to Hartl (1996), stress proteins do not contain steric information specifying correct folding; instead, they prevent incorrect interactions within and between nonnative polypeptides, thus typically increasing the yield but not the rate of folding reactions, as do folding catalysts, such as protein disulfide isomerases (PDIs) and peptidyl-prolyl isomerases (PPIs). Moreover, Hartl (1996) believes that each folding step is preceded by a release of the polypeptide from the sp70 complex, providing the option for the substrate of either folding, rebinding, or being transferred to another chaperone

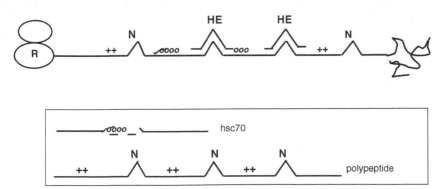

Figure 3.1 Schematic drawing of the assistance of HSCs in the folding and assembling of polypeptides leaving ribosomes (R). The polypeptide contains multiple binding sites with positively charged sequences and neighboring neutral (N) residues. The HSC contains negatively charged residues on an α-helix structure. Ionic attraction leads to a hydrophobic enclosure (HE) of the neutral residues of the polypeptide and a transient binding. Release and subsequent folding is considered a stepwise process.

system. It should be noted that an sp70 complex is defined as the combination of at least one sp70 and a smaller stress protein. The binding and release cycle of the sp70 is modulated by the smaller stress protein (Hartl, 1996).

The existence of several binding sites on the nascent polypeptide suggests the binding of more than one sp70 molecule to one polypeptide chain (Rothman, 1989). The "folding models" of Rothman (1989) and Hartl (1996) do not differ much: Rothman (1989) emphasizes the catalytic action of sp70, whereas Hartl (1996) emphasizes the stabilizing role of sp70 with transient interactions between it and the polypeptide. The latter model is preferred because it was recently found that the energy from ATP hydrolysis used in the folding process is not required for binding and dissociation, but for conformational changes in the stress protein. This will be discussed further below.

Chaperoning precursor proteins across intracellular membranes – Cytoplasmic cognates chaperone polypeptides on their way to and through the membranes of mitochondria and the endoplasmic reticulum (ER). To enable the protein to pass the membrane, the cytoplasmic cognates keep it in an unfolded or partially folded state. The precursors are provided with signal sequences, which are proteolytically removed after membrane passage (Ellis and van der Vies, 1991). Membrane passage through the mitochondrial membrane is illustrated in Figure 3.2.

Figure 3.2 shows that membrane receptors are involved in membrane translocation. Membrane passage is mediated by mitochondrial hsp70, the stress protein that is located in the matrix of the mitochondrium, and by an energized inner membrane exerting an electrophoretic effect on the positively charged precursor (Bauer et al., 2000; Langer and Neupert, 1994; Kunau et al., 2001).

In all the functions described for the members of the sp70 family, ATP binding and hydrolysis to adenosine diphosphate (ADP) is involved, providing the energy needed for the conformational changes in the stress proteins. Most of the hypotheses that emerged in the 1980s concerning the role of ATP hydrolysis suggested that hydrolysis was necessary for release of the peptide (Hightower et al., 1994). Palleros et al. (1993) reported that release of unfolded proteins preceded ATP hydrolysis. A proposed model is shown in Figure 3.3.

Figure 3.3 shows that there are four steps in peptide binding and release: peptide binding, ATP/ADP exchange, peptide release, and ATP hydrolysis. Remembering that the ATP/ADP domain is at the N-terminal side of the stress protein, while the peptide-binding site is at the C-terminal side, one can see that an action initiated on one side of the stress protein delivers a response on the other side. Thus, the binding and ATPase cycles work as a "click system" (Hightower et al., 1994). Note the difference between ATP binding and phosphorylation. In the phosphorylation, phosphate is bound to some amino acids (mostly serine or threonine) of a protein by hydrolysis of ATP to ADP. Proteins can be phosphorylated in response to a wide variety of stimuli, leading to activation (or deactivation). By stepwise

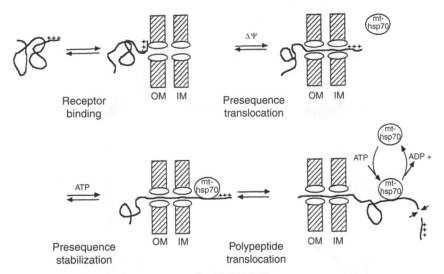

Figure 3.2 Model of mitochondrial hsp70-mediated membrane translocation of mitochondrial precursor proteins. The precursor binds to membrane receptors. Membrane passage through the outer and inner membrane is mediated by mitochondrial hsp70 (hsc71) by activating the membrane receptor. Abbreviations: mt-hsp70 = mitochondrial hsp70; OM = outer membrane; IM = inner membrane; $\Delta\psi$ = electrophoretic force; ATP = adenosine triphosphate; ADP = adenosine diphosphate. (After Langer, T. and Neupert, W., in *The Biology of Heat Shock Proteins and Molecular Chaperones*, Morimoto, R.I. et al., Eds., Cold Spring Harbor Laboratory Press, Cold Spring Harbor, New York, 1994. With permission.)

phosphorylation, the level of activation and response can be modulated (Arrigo and Landry, 1994). The role of phosphorylation in stress protein gene expression will be discussed further in Section 3.3.

3.2.1.2   The mitochondrial sp70: hsc71

Just like 95% of the other mitochondrial proteins (Pfanner, 1990), hsc71 is encoded in the nucleus, synthesized in the cytoplasm, and translocated to the matrix of the mitochondrium, where it matures to fulfill its role of assisting in the folding and assembly of mitochondrial proteins (see Table 3.1). The messenger RNA (mRNA) of hsc71 contains three introns (Rubin et al., 1993), and it is therefore questionable whether it can be processed under stressful conditions (splicing of introns will be discussed further in Section 3.3). The assembly of oligomeric proteins to functional protein complexes in mitochondria takes place in the presence of "chaperonins" (Gutenby et al., 1990). In mammalian cells, this assembly is performed by hsp58, which belongs to the sp60 family (Welch, 1992). The corresponding type in *Drosophila* species is not yet known. The chaperonins are constitutively present and induced by stress (Gutenby et al., 1990).

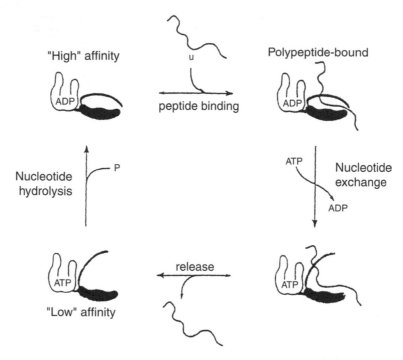

Figure 3.3 Model for the reaction cycle of an hsp70 protein in relation to the ATP/ADP cycle. The ADP-bound state favors polypeptide binding, the ATP-bound state favors polypeptide release, and the hydrolysis of ATP returns the component to the high-affinity ADP state. Abbreviations: ATP = adenosine triphosphate; ADP = adenosine diphosphate; u = unfolded protein (After Craig, E.A. et al., *Cell* 78: 365–372, 1994. With permission.)

3.2.1.3 *The sp70 in the ER: hsc72*

Hsc72 associates tightly in the ER with newly synthesized proteins that are incompletely assembled, have mutant structures, or are incorrectly glycosylated. Hsc72 has been suggested to prevent secretion from the ER of incorrectly folded or incompletely assembled proteins, to promote folding or assembly of proteins, or to solubilize protein aggregates within the ER lumen (Kassenbrock et al., 1988; Lindquist and Craig, 1988). Moreover, hsc72 may facilitate import of precursor proteins in the same way as hsc71 in mitochondria, except for the translocation over the inner membrane. It seems, however, that hsc72 is not always required for import of precursor proteins into the ER. For instance, in special cases when import takes place directly from the ribosome on the outside of the rough endoplasmic reticulum (RER) (cotranslational translocation), hsc72 would not be absolutely required (Brodsky and Schekman, 1994). Hsc72 is encoded in the nucleus, yielding an mRNA with two introns. Activation of the gene expressing hsc72 requires a signal from the ER. For that purpose, there is a special signal transduction pathway that will be discussed in Section 3.4. The precursor protein of hsc72 has a hydrophobic leader at the N-terminal for import into the ER, while the C-terminal has a retrieval sequence so that secretion from the ER is prevented (Rubin et al., 1993).

Unlike other cellular organelles, the ER has an oxidizing environment, which promotes the formation of disulfide bonds between cysteine residues. Protein disulfide isomerase (PDI), a multifunctional resident ER protein, catalyzes the formation and isomerization of disulfide bonds (Luz and Lennarz, 1996; Wei and Hendershot, 1996).

Calcium, as second messenger, is stored in the smooth endoplasmic reticulum (SER) of the ER. To maintain the redox state with an oxidizing environment, necessary for the correct assembly of proteins in the ER, calcium is retained by calreticulin (Wei and Hendershot, 1996). According to Li and Lee (1991), depletion of sequestered Ca^{2+} in the ER also blocks protein glycosylation, resulting in the induction of the stress protein of the ER. This is coupled with the repression of hsp70. The role of calcium in relation to the induction of stress proteins will be discussed further in Chapter 4.

According to Rubin et al. (1993), hsc72 levels in *Drosophila* do not increase upon heat shock. This is not only due to the fact that this stress protein contains two introns, but is also due to limited gene-expression potential. Further reference to this aspect will be made in the Sections 3.3 and 3.4. The stress protein in the human ER, grp78 (identical to "binding protein" BiP; see Table 3.1) is slightly inducible by heat and several other stressors, including glucose deprivation (Gething et al., 1994; Watowich and Morimoto, 1988). Some of the stress proteins in human cells (grp75 in the mitochondrium and grp78 in the ER) are glucose regulated, and therefore called glucose regulated proteins (GRPs). When induced by glucose deprivation, other stress proteins seem to be repressed (Sanders, 1990; Watowich and Morrimoto, 1988). The effect of induction of GRPs and the repression of other stress proteins is also observed when the cell is deprived of oxygen (Sanders, 1990). This is an interesting aspect from a control point of view and will be discussed further in Section 3.4. It is not yet clear whether some of the members of the sp70 family of *Drosophila* are also glucose regulated, but hsp83 of the sp90 family in *Drosophila* is (see Section 3.2.2).

3.2.1.4 The inducible cytoplasmic (nuclear) stress proteins: hsp70 and hsp68

As Table 3.1 shows, hsp70, the major stress-inducible protein, is encoded in *D. melanogaster* by several homologous genes located on two chromosomal sites (87A7 and 87C1). The latter locus contains three to five copies of the hsp70 gene, depending on the fly stock or tissue culture cell. These copies are almost identical, suggesting a high rate of intralocus gene conversion (Pauli and Tissières, 1990). Hsp68 is homologous to hsp70, with about 15% divergence at the DNA sequence level (Pauli and Tissières, 1990).

Under stressful conditions, the inducible stress proteins of the sp70 family are quickly and abundantly induced. This takes a matter of minutes rather than hours; for example, in *D. melanogaster*, the increase in message after a heat shock is noticed after 4 min (Lindquist, 1986). In this species, hsp70 and hsp68 are most abundantly induced, but their levels after induction still do not reach the level of the constitutively present hsc70 (Lindquist and Craig, 1988; Pauli and Tissières, 1990). It is suggested that, during stress, stress proteins protect essential proteins against denaturation and repair damage and are possibly involved in disassembly and degradation of abnormally folded proteins (Hightower et al., 1994; Welch, 1992). For that purpose, under stressful conditions, hsp70 in, for example, *D. melanogaster* is concentrated in the nucleus and on cell membranes and accumulated in the nucleolus (Lindquist and Craig, 1988). Immuno-electron microscopy shows that hsp70 is directly associated with damaged preribosomes (Pelham, 1990), supporting the opinion that maintaining ribosomal integrity is one of the most important steps in defense against stress (Welch, 1990). Hartl (1996) states that, under stressful conditions, hsp70 exerts its protective action in mammalian cells, in cooperation with hsp40 and hsp90, complemented by a class of ATP-independent SHSPs of about 25 kDa, which may act as a first line of defense against aggregation.

During recovery, hsp70 is translocated to the cytoplasm (Lindquist and Craig, 1988), where it participates in the degradation of damaged proteins. Its regulation will be described further in Section 3.3. It is likely that other members of the sp70 family fulfill similar roles during stress in the other compartments (Pelham, 1990). Thus, during stress, hsc72 may bind to secretory proteins in the ER, normal glycosylation of which, possibly owing to glucose starvation, has been interrupted, leading to misfolding and precipitation (Pelham, 1990). Similarly, hsc70, constitutively present in the cytoplasm, is (partly) translocated to the nucleus during stress to assist hsp70 in its activities. This and the fact that the homology between the amino acid sequences of hsc70 and hsp70 is high (74%) could give the impression that the differences between the two main types of stress proteins in the cytoplasm and nucleus are more determined by differences in expression and regulation than by structure and function. It should be realized, however, that in *Drosophila* under unstressed conditions, hsp70 is virtually undetectable and that this protein does not assist hsc70 in its tasks at normal growth and development, as discussed in Section 3.2.2.1. Beyond that, Parsell and Lindquist (1994) demonstrated that expression of hsp70 in *Drosophila* under normal conditions is detrimental to growth and cell division. In contrast, in both yeast and mammalian cells, heat-inducible hsp70 proteins are expressed at a substantial level at normal temperatures under unstressed conditions (Feder et al., 1992).

3.2.2   hsp83

We know that hsp83 is expressed by a single gene in *D. melanogaster*, located on the chromosomal site 63BC (see Table 3.1). The transcription unit is interrupted by one intron of variable size depending on the fly strain. It separates the 5'-untranslated leader from the coding region. The inhibition of splicing of this intron at high temperatures (severe heat shock) accounts for the relatively low optimum of 33°C of expression of hsp83 (normal optimum is 36 to 37°C for stress proteins without introns in *Drosophila*) (Pauli and Tissières, 1990) (see Section 3.3 for further discussion). Apart from hsp83, the cytoplasmic hsc70b, the mitochondrial hsc71 and hsc72 in the ER have one or more introns, whereas the cytoplasmic hsc70 and the inducible proteins, hsp70 and hsp68, do not have an intron.

A soluble cytoplasmic protein, hsp83 is methylated (at lysine and arginine residues) and phosphorylated (at serine and threonine residues) (Lindquist, 1986). It is constitutively present and induced by stressors, such as heat and glucose deprivation (Lindquist, 1986). This stress protein in *Drosophila* therefore is glucose regulated. Its presence is abundant but life-stage dependent and tissue specific. For example, hsp83 is developmentally induced during oogenesis in *D. melanogaster* (Lindquist and Craig, 1988). Moreover, in cells that respond to steroid hormones, hsp83 binds to uncomplexed receptors (Plesofsky-Vig, 1996). When a ligand binds to the receptor, the hsp83 dissociates and the receptor is transformed and becomes biologically active, leading to gene transcription (Welch, 1990). This regulation is illustrated in Figure 3.4.

Stepanova et al. (1996) noticed that in accordance with its role in stabilizing hormone receptors, in insect cells hsp83 also exerts a stabilizing role to CdK4, a cyclin-dependent protein kinase, activating cyclins in cell cycle steps. Mosser et al. (1993) assigned a stabilizing role to hsp90 in human cells (equivalent to hsp83 in *Drosophila* cells) to maintain transcription factors in an inactive form (see Section 3.3 for further discussion). Other findings suggest that the HSP may shuttle important membrane proteins around the cell, holding them in an inactive form during transport. For instance, Rutherford and Zuker (1994) reported on the association of the tyrosine kinase, pp60 v-stc, with an hsp90 heterocomplex. Activation of this enzyme is correlated with its release from the complex and translocation into the plasma membrane. Apart from that, Rutherford and Zuker (1994) noticed that heat shock causes dissociation of hsp90 from the receptor. Craig et al. (1994) reported on *in vivo* studies showing that hsp90 acts only on specific protein targets, binding to many steroid hormone receptors, kinases, and calmodulin, all involved with signal transduction. Bensaude et al. (1996) reported that triggering enzymes upstream in the mitogen-activated protein kinase (MAPK) cascade form heterocomplexes with hsp90 when inactive and dissociate from it upon activation, for example, by heat. This also illustrates the involvement of constitutively present stress proteins in the transduction pathways that transduce cellular signals, including stress signals.

These observations led to the postulation by Bohen and Yamamoto (1994) that HSPs may exert regulatory effects on a broad range of intracellular signaling pathways and thus may serve as a common link between these systems. We know that hsp83 is induced by stress. According to Hartl (1996), it then assists hsp70 in protecting native proteins against denaturation.

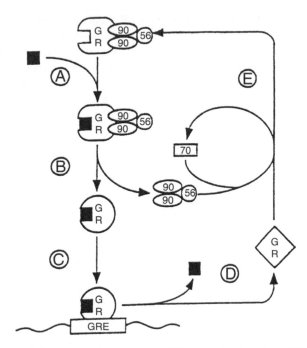

Figure 3.4 Cycle of glucocorticoid receptor (GR) activation by ligand binding (A) and dissociation of a stress protein complex (B), followed by activation of transcription by binding of the receptor-ligand complex to a glucocorticoid response element (GRE) in the DNA leader for expressing the target protein (C). Dissociation of the receptor-ligand complex from the GRE leads to reconstitution of the inactive GR complex (D) by binding of the stress protein complex to the GR (E), assisted by hsp70 in a stabilizing role. The stress protein complex consists of a dimer of hsp90 (hsp83 in the case of *Drosophila*) and an auxiliary protein, hsp56. The ligand is marked as a black square. (After Bohen, S.P. and Yamamoto, K.R., in *The Biology of Heat Shock Proteins and Molecular Chaperones*, Morimoto, R.I. et al., Eds., Cold Spring Harbor Laboratory Press, Cold Spring Harbor, New York, 1994. With permission.)

3.2.3   Small HSPs

Four small heat shock proteins (SHSPs) of low molecular weights of 22, 23, 26, and 27 kDa (Tanguay et al., 1999) have been identified in *Drosophila* (see Table 3.1), although both the size and number may vary among species (Lindquist, 1986). The four SHSPs of *D. melanogaster* have an overall homology of about 50% (Lindquist, 1986), so they form a rather heterogenous group. Each member of this class of SHSPs in *D. melanogaster* is encoded by one gene, and these four genes are clustered on locus 67BC (Pauli and Tissières, 1990). SHSPs have two major domains. The N-terminal domain is only moderately conserved between SHAPs of the same species (Arrigo and Landry, 1994), but some parts in the most important regions of this domain are well conserved. For example, 3 of the 4 *Drosophila* SHSPs share a region comprising the first 15 N-terminal amino acids that resembles signal peptides

(Arrigo and Landry, 1994). This region is very hydrophobic and might allow some interaction with membranes (Pauli and Tissières, 1990; Tanguay et al., 1999). A domain in the second half of the polypeptide is homologous to a mammalian carboxy-terminal region of the α-crystalline polypeptides, which are the major structural proteins of the lens (Pauli and Tissières, 1990; Tanguay et al., 1999). The homology with α-crystalline may relate to the capacity of SHSPs to form higher-order structures (super-aggregates) (Arrigo and Landry, 1994).

Cytoplasmic particles containing the SHSPs, complexed with the SHSP RNAs, have been isolated from *Drosophila* after recovery from heat shock (Lindquist, 1986). The SHSPs show a temporal and spatial expression during normal development of the fly, during which the genes on locus 67BC are commonly expressed, but tissue-specific, resulting in concentrations that may differ 100-fold (Pauli and Tissières, 1990). The fluctuations during development seem to be hormonally controlled. In *D. melanogaster,* expression of SHSPs seems to be controlled by the moulting hormone, "ecdysone" (Welch, 1990). Hormonal induction during development is regulated by steroid receptor–binding sequences, located far upstream of the beginning of the transcription of the SHSP gene (Arrigo and Landry, 1994). Recent data further suggest that, under normal conditions, SHSPs have a homeostatic function at the level of signal transduction (Arrigo and Landry, 1994). In this context, it seems logical that small HSPs have also been reported to act as suppressors of programmed cell death (apoptosis) and to act through modulation of the cell redox (Tanguay et al., 1999) (see Chapter 8 for further discussion). In search of proteins that interact with SHSPs, Tanguay et al. (1999) reported that hsp27 binds in the nucleus a ubiquitin-conjugating enzyme that is involved in cellular proteolytic activities (see Section 3.2.4). This led them to suggest that SHSPs may play a role in the proteolytic process.

Under stressful conditions, the SHSPs are strongly induced in large quantities in every cell (Arrigo and Landry, 1994), where they are found within and around the nucleus, associated with RNA (Lindquist, 1986) and microfilaments, and forming super-aggregates (Arrigo and Landry, 1994). Arrigo and Landry (1994) also believe that thermal protection is achieved by stabilizing the microfilaments. In *D. melanogaster,* this protection is performed by hsp27. In contrast to mammalian hsp27, the protein in *D. melanogaster* is mainly localized in the nucleus, leading to different protection activities. It was observed, for instance, that the *Drosophila* hsp27 effectively protects against inhibition of protein synthesis during heat shock but was excluded from interactions with cytoplasmic actin, to prevent collapse of the microfilaments in the cytoplasm (Arrigo and Landry, 1994). Under normal conditions, however, localization of hsp27 seems to be dynamic; for example, during the early stages of oogenesis it is localized in the nucleus, but from stage 8 on it relocalizes to the cytoplasm (Tanguay et al., 1999).

One intriguing aspect of SHSP biochemistry is the ability of many of these proteins to become phosphorylated in response to a large variety of stimuli, suggesting that, in addition to the level of expression, the level of phosphorylation can modulate their function in cellular physiology. These stimuli include toxic agents and treatments, such as heat shock, arsenite, and hydrogen peroxide; mitogens and differentiation factors, such as thrombin; growth factors; and calcium ionophores (Arrigo and Landry, 1994). Two of the *Drosophila* SHSPs are also phosphorylated in response to ecdysterone. Notwithstanding the fact that phosphorylation is detectable within a few minutes, it does not activate SHSP gene transcription and the accumulation of the protein. It mostly affects preexisting proteins. Figure 3.5 shows the time course of phosphorylation and heat shock–induced protein synthesis of hsp27 in Chinese hamster.

Activation of hsp27 kinase activity by external stimuli is the major mechanism regulating phosphorylation of shsp27. Oxyradical generating agents are strong inducers of hsp27 kinase. Moreover, oxyradical-sensitive phosphatases may probably be inactivated. Both effects may lead to phosphorylation of hsp27. An upstream activator of hsp27 kinase may be the mitogen-activated protein kinases (MAPK). All known inducers of hsp27 phosphorylation, including heat shock and hydrogen peroxide, also induce MAP kinase activity (Arrigo and Landry, 1994). However, scavenging of free radicals reduced phosphorylation levels of SHSP, but this treatment did not affect phosphorylation levels induced by heat shock, suggesting that pathways independent of oxyradicals also exist (Arrigo and Landry, 1994).

Together, the function and regulation of SHSPs are complicated. Perhaps constitutively present SHSPs are activated by phosphorylation via the MAPK

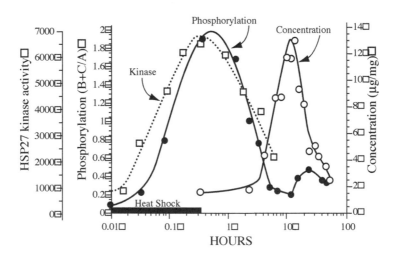

Figure 3.5 Heat-shock-induced hsp27 accumulation and phosphorylation and hsp27 kinase activity in Chinese hamster. The figure shows that phosphorylation precedes protein synthesis. (After Arrigo, A.P. and Landry, J., in *The Biology of Heat Shock Proteins and Molecular Chaperones*, Morimoto, R.I. et al., Eds., Cold Spring Harbor Laboratory Press, Cold Spring Harbor, New York, 1994. With permission.)

cascade. However, SHSPs that are strongly induced under stressful conditions probably will form large unphosphorylated aggregates. In the former case, the HSPs may fulfill their role in protecting and chaperoning special proteins and enzymes. In the latter case, they may protect nuclear structures including RNA. Their role in supporting the cellular redox state and preventing apoptosis will be discussed in Chapter 8.

3.2.4   Ubiquitin

Ubiquitin, as its name implies, is found abundantly in all eukaryotic cells (Lindquist and Craig, 1988). It is a compact globular protein consisting of 76 amino acids. It is extremely well conserved: Only three amino acids differ between yeast and humans (Schlesinger, 1990). There are two types of genes encoding ubiquitin and both have unusual sequences in their reading frames, leading to expression of polyubiquitin and ubiquitin linked to ribosomal proteins. There are several genes for both types, one of which is heat inducible (Schlesinger, 1990). The roles of ubiquitin are shown in Figure 3.6.

The functions of ubiquitin shown in Figure 3.6 can be described as follows:

1. Ubiquitination of histones: The exact function is unknown, but it seems to be essential for DNA to be active in the interphase of the cell cycle, so that transcription or replication is made possible. During cell division, condensed chromatin is devoid of ubiquitinated histones.
2. Protein degradation: In this pathway, ubiquitin acts as a factor for cytoplasmic ATP-dependent protein degradation. This proteolytic degradation system normally functions to "turn over" cytoplasmic

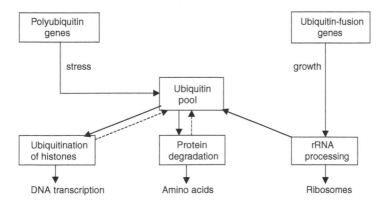

Figure 3.6 Role of ubiquitin in cell metabolism. Under normal conditions, the ubiquitin-fusion gene provides ubiquitin for cell metabolism. Under stressful conditions, the polyubiquitin gene is activated for the same processes. Abbreviation: rRNA = ribosomal RNA (Adapted from Schlesinger, M.J., in *Stress Proteins: Induction and Function*, Schlesinger, M.J. et al., Eds., Springer-Verlag, Heidelberg, 1990. With permission.)

and nuclear proteins. Under stressful conditions, the ubiquitin concentration is increased threefold to ubiquitinate aberrant proteins and mark them for proteolysis (Schlesinger, 1990).

3. Ribosomal RNA (rRNA) processing: Ubiquitin-RNA is processed together with the RNA of two ribosomal proteins. The ubiquitin moiety of the fusion protein is postulated to stabilize the ribosomal protein during its synthesis and transport it to its site of assembly in the nucleolus (Schlesinger, 1990).

All three functions of ubiquitin are also important under stressful conditions, but two important changes affect them:

1. In heat-shocked cells, ubiquitinated histones disappear rapidly, leading to condensation of chromatin and decreased DNA processing. Ubiquitin hydrolase, which removes ubiquitin from histones, may be activated by stress, and the released ubiquitin could exit the nucleus and participate in the ubiquitin-dependent proteolytic degradation pathway (Schlesinger, 1990). Thus, under stressful conditions, the cell gives preference to ubiquitin in the degradation role. In this way, new building blocks (amino acids) for synthesis of stress proteins become available. The important route of DNA processing is suppressed, however.
2. The fusion gene is the major *de novo* source of free ubiquitin in growing cells, but this source becomes limiting under stressful conditions, and additional ubiquitin is obtained by activation of the polyubiquitin gene (Schlesinger, 1990). Bell et al. (1988) investigated the effect of heat shock on ribosome synthesis in *D. melanogaster* and observed that transcription of rRNA precursor continued but that this rRNA was rapidly degraded and subsequent ribosome synthesis was halted. This may be a consequence of impaired functioning of spliceosomes during heat stress.

In summary, under stressful conditions, preference is given to the degradation pathway, while the other two important routes, of DNA processing and ribosome synthesis, are suppressed.

3.3 Gene expression, posttranscriptional, and translational regulation of stress proteins

3.3.1 Gene expression of stress proteins

The key to the pathway to gene expression of stress proteins, both constitutively present and induced under stressful conditions, is transcription factors, commonly termed heat-shock (transcription) factors (HSFs). For the sake of simplicity the latter terminology will be followed.

The general consensus on the sequence of events is that there exists an equilibrium between constitutively present HSFs and excess stress proteins that transiently and loosely bind HSFs. Because the inducible hsp70 is absent in *Drosophila* under normal conditions (Feder et al., 1992; Krebs and Feder, 1997a), it is believed that the HSF-binding stress protein will be predominantly the constitutively present hsc70. Owing to a stress situation, the stress proteins will bind to their target proteins, either aberrant or essential, and abolish their interactions with HSFs. The HSFs become activated and bind to upstream DNA sequences of the stress protein genes, called heat-shock elements (HSEs), thus inducing transcriptional activation of the stress protein genes. This leads to mRNA transcripts and the synthesis of stress proteins. This sequence of events is illustrated in Figure 3.7

Gene expression will be discussed in three parts: the HSF, the heat-shock gene promoter, and gene expression and its regulation.

3.3.1.1   The heat-shock (transcription) factor (HSF)

In yeast and *Drosophila* only one HSF has been identified. However, in larger eukaryotes more HSF types are present: In humans two HSFs (HSF1 and 2)

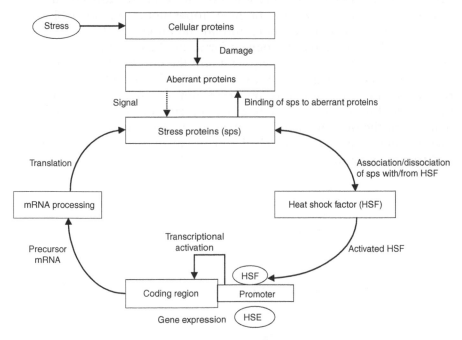

Figure 3.7 Sequence of events in the cell under stress. Stress causing protein damage leads to accumulation of aberrant proteins, which exert a greater affinity on stress proteins than HSF molecules do. Stress proteins consequently dissociate from HSF, which then becomes activated. Activated HSF molecules bind to heat-shock elements (HSEs) on the promoter of the gene expressing the transcripts, leading to mRNA processing and synthesis of new stress proteins. Abbreviations: HSF = heat-shock factor; mRNA = messenger RNA; SP = stress protein.

and in chicken, at least three. The *Drosophila* protein is dispensable for cell growth and viability but is required for oogenesis and early larval development. These two developmental functions appear not to be mediated through the induction of HSPs, implicating an action unrelated to its characteristic function (Jedlicka et al., 1997). The human HSF1 is the functional homolog to the *Drosophila* HSF and is activated *in vivo* by heat and numerous other forms of physiological stress. The human HSF2 is developmentally activated and acquires DNA-binding activity after treatment with hemin that contains growth factors (Sarge et al., 1993). HSF2 is not stress responsive, but under certain conditions, HSF1 and 2 can be co-activated, binding the same gene and activating it synergistically (Morimoto et al., 1994b). Figure 3.8 compares the human HSF1, the *Drosophila* HSF, and the yeast HSF.

Figure 3.8 shows that the yeast HSF is most divergent. The *Drosophila* HSF contains a DNA-binding domain and two leucine-zipper domains. Analysis of the DNA-binding domain shows an α-helical structure with the putative helix-turn-helix motif (Craig et al., 1994; Wu et al., 1994). The leucine-zipper domains consist of a leucine-rich region in which successive leucine residues occur every seventh amino acid, forming an α-helical structure in which adjacent leucine residues occur every two turns on the same side of the helix. In this way two or three zipper motifs can form HSF dimers or trimers in a coiled motif. Under normal conditions, the *Drosophila* HSF is constitutively present in a monomeric form, and this form is maintained by an interaction between leucine zipper 4 and the domain containing leucine zippers 1, 2, and 3 (Figure 3.8). Upon heat shock, this interaction is interrupted and the HSF molecule is converted from a monomer to a homotrimer, the activated form for DNA binding. Trimerization takes place between the domains of the three monomers containing the leucine zippers 1, 2, and 3. These domains are therefore called trimerization domains. The monomer and trimer configuration is shown in Figure 3.9.

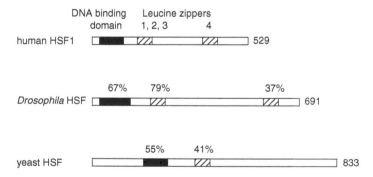

Figure 3.8 Comparison of human heat-shock factor 1 (HSF1), *Drosophila* HSF, and yeast HSF, with structural homology% between conserved domains (After Wu, C., et al., in *The Biology of Heat Shock Proteins and Molecular Chaperones,* Morimoto, R.I. et al., Eds., Cold Spring Harbor Laboratory Press, Cold Spring Harbor, New York, 1994. With permission.)

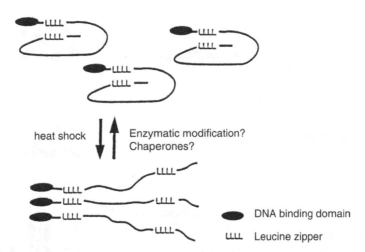

heat shock

Enzymatic modification?
Chaperones?

LLLL   DNA binding domain

LLLL   Leucine zipper

Figure 3.9 Heat-shock factor (HSF) trimerization during heat shock (After Wu, C. et al., in *The Biology of Heat Shock Proteins and Molecular Chaperones*, Morimoto, R.I. et al., Eds., Cold Spring Harbor Laboratory Press, Cold Spring Harbor, New York, 1994. With permission.)

As expected for an active DNA-binding protein, the *Drosophila* HSF in trimeric form is localized in the nucleus after heat shock. The monomer present in unshocked *Drosophila* cells is also localized in the nucleus. This finding was unexpected in view of previous biochemical studies that placed the HSF in the cytoplasm (Mosser et al., 1993; Wu et al., 1994). Furthermore, the monomer is found to be diffusely located over the chromatin in the absence of heat shock. After heat shock, the diffused pattern is replaced by discrete staining at more than 200 distinct cytological loci. The loci include the major puff sites, as well as major developmental loci, which could reflect an alternate function for the HSF trimers of *Drosophila* in the repression of normal gene expression during heat shock (Wu et al., 1994).

In budding yeast, the HSF is not stored in a monomeric form but is already bound to the HSE in DNA in its trimeric form. This is why the yeast HSF in Figure 3.8 does not contain a leucine-zipper 4 domain.

Upon heat shock, HSE becomes associated with the activated HSF, with occupancy correlating closely with the levels of HSF DNA-binding activity. Likewise, during gradual attenuation, occupancy decreases and, correspondingly, the HSEs are no longer occupied by HSF. Half-life tests suggest that the activated form of HSF may be modified during the recovery phase, so as to interfere with its DNA-binding activity and the conversion of the active trimer to the inert monomeric form. It has been speculated from early studies with *Drosophila* that HSPs are possible candidates for keeping HSFs in this inactive state. This is considered to be a negative control, that is, kept in a monomeric inactive form. It is believed, however, that stress proteins associating with the HSFs have more of a regulatory role than a stabilizing role. This leads to the hypothesis that there is a critical balance between HSF and

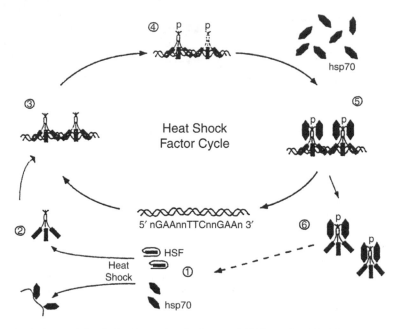

Figure 3.10 A model of the heat-shock factor (HSF) cycle: (1) Monomeric state, (2) trimeric state, (3) DNA binding, (4) phosphorylation step, (5) transcription, (6) dissociation. Abbreviation: hsp = heat-shock protein (After Morimoto, R.I. et al., in *The Biology of Heat Shock Proteins and Molecular Chaperones*, Morimoto, R.I. et al., Eds., Cold Spring Harbor Laboratory Press, Cold Spring Harbor, New York, 1994b. With permission.)

its negative regulatory molecules that maintains HSF in either the non-DNA-binding state or the DNA-binding state (Morimoto, 1993; Morimoto et al., 1994b; Mosser et al., 1993; Shi et al., 1998). This subject is discussed further in Sections 3.3.1.3 and 3.3.3. Figure 3.10 shows a model for DNA-binding of HSF.

The steps shown in Figure 3.10 represent three stages: HSF trimerization and DNA binding, transactivation and transcription, and dissociation and HSF deactivation. The stage including transactivation of the stress protein gene and leading to transcription requires phosphorylation of the HSF molecule after binding to the DNA promoter of the stress protein gene. Phosphorylation takes place at serine and threonine residues, thereby unmasking a so-called transactivation domain at the C-terminal side of the HSF (Zhong et al., 1998). Such phosphorylation would bring about conformational changes in the HSF molecule, allowing it to interact with components of the constitutive transcriptional apparatus and to activate transcription of the stress protein gene (Latchman, 1995). *Drosophila* HSF also contains transactivation domains (Wu, 1995) and is phosphorylated after DNA binding to activate transcription (Latchman, 1995; Zhong et al., 1998). Mager and De

Kruijff (1995) reported, however, on an alternative opinion that phosphorylation of HSF might occur as a consequence of the formation of a transcription initiation complex rather than as a prerequisite for this formation.

Whatever the sequence of events appears to be, Xia and Voellmy (1997) obtained evidence that phosphorylation of HSF1 trimers increases their transcriptional activity. Additional proof resulted from tests by Winegarden et al. (1996). By exposing *Drosophila* cells to sodium salicylate, they observed HSF–DNA binding but no hsp70 gene transcription. It appeared that phosphorylation of this complex was not accomplished, probably owing to a decrease of cellular ATP levels, observed at the same time as HSF activation. The results were concentration dependent, however. At very high concentrations of sodium salicylate, even heat-induced gene transcription was inhibited, and the ATP level was reduced to one third of normal. According to Winegarden et al. (1996), a shortage of ATP caused by sodium salicylate could also interfere with the chaperone function of heat-shock proteins, leading to accumulation of aberrant proteins and subsequent activation of HSF–DNA binding.

3.3.1.2   The heat-shock gene promoter

Almost all promoters of heat-shock genes expressing HSPs in different organisms have HSEs containing the 5-bp unit 5'-GAA-3', where each repeat is inverted relative to the immediately adjacent repeat, leading to a dyad 5'-GAA — TTC-3' or even a contiguous array (Fernandes et al., 1994; Lis et al., 1990). For example, the promoter region of the *Drosophila* hsp70 gene has four HSEs, each containing four or five 5-bp units. Figure 3.11 compares the number and locations of HSEs in hsp70, hsp83, and hsp27 genes.

As a rule, HSP gene promoters contain multiple HSEs, each with three to six 5-bp units. The hsp83 promoter contains only one HSE, but with seven or eight 5-bp units, depending on the *Drosophila* species. An arrangement of it is shown in Figure 3.12.

It should be noted that the hsp83 gene is also glucose regulated (see Section 3.2.2), but it is not yet clear whether its promoter contains an unfolded protein response element (UPRE) (see Section 3.4). Induction of the hsp83 gene at glucose deprivation might also be regulated via basal signaling pathways, involved in the conversion of glycogen into glucose.

The localization within the promoter is another variable and may vary considerably, as can be seen from Figure 3.11. The multiple 5-bp repeats of the HSE are positioned on the DNA double helix with approximately three-fold symmetry. In this way, the DNA-binding domains of the trimeric HSF fit into the HSE of the DNA (Craig et al., 1994; Lis and Wu, 1993). The HSF trimer displays a remarkable flexibility in its ability to interact with HSEs containing different numbers and arrangements of 5-bp units. The smallest array that shows detectable binding of *Drosophila* HSF contains two 5-bp units, but a complete minimal binding site for trimeric HSF contains three 5-bp units. The binding affinity of HSF in *Drosophila* increases with the number of 5-bp units, indicating a cooperative binding (Sewell et al., 1995).

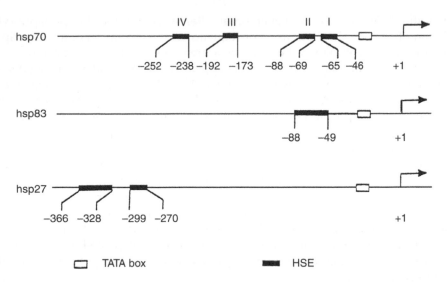

| | TATA box | | HSE |

Figure 3.11 Heat-shock gene promoter and regulatory regions of the *Drosophila* hsp70, hsp83 and hsp27 genes. Abbreviations: hsp = gene for expression of heat-shock proteins. (After Fernandes, M. et al., in *The Biology of Heat Shock Proteins and Molecular Chaperones,* Morimoto, R.I. et al., Eds., Cold Spring Harbor Laboratory Press, Cold Spring Harbor, New York, 1994. With permission.)

The resulting range in affinities for HSF by the various stress protein genes may account in part for their differential response to heat shock (Fernandes et al., 1994). Multiple HSEs also lead to binding of multiple HSFs, which can reinforce each other's effects, stimulating transcription (Lis et al., 1990). This model also supports the opinion that transcription is quantitatively regulated (Harshman and James, 1998), that is, proportional to the number of 5-bp units (Lis et al., 1990) and the concentration of activated HSF. HSC3, the *Drosophila* gene expressing hsc72 for the ER, contains a single pair of inverted units, which is believed to be insufficient for a stress-induced activation of transcription (Rubin et al., 1993). These data are consistent with the statement made in Section 3.2.1.3 concerning inducibility of hsc72 under stressful conditions.

3.3.1.3   Gene expression and regulation

In uninduced *Drosophila* cells, heat-shock promoters are poised for a rapid change in transcription. While the bulk of the DNA is associated with histone proteins and the formation of a tightly packed chromatin structure, the binding of a GAGA factor to the heat shock–gene promoters results in displacement of the histone-containing nucleosomes from the promoter region. Furthermore, prior to heat shock, the TATA-box binding protein (TBP) of the TATA-box binding protein complex (TFIID) is bound to the TATA-box (Latchman, 1995; Lis and Wu, 1993). This is illustrated in Figure 3.13.

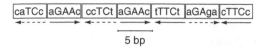

Figure 3.12 The arrangement of the 5-bp units of the heat-shock element (HSE) in the hsp83 gene promoter of a *Drosophila* species. Arrows indicate the orientations of the units, and dashed lines denote imperfect matches to the consensus nGAAn sequence. (After Sorger, P.K., *Cell* 65: 363–366, 1991. With permission.)

Moreover, RNA polymerase II is also associated with the uninduced heat shock–gene promoter, located in the region between –12 and +65. The presence of polymerase II at the 5′ end of uninduced heat-shock genes suggests that these genes can respond rapidly to a heat shock or other stress. When the gene is transcriptionally activated, the paused polymerase II starts elongating from the promoter within 30 to 60 seconds, reaching the 3′ end of the gene within 120 seconds (Fernandes et al., 1994). When cells are stimulated by heat shock, heat shock–gene promoters are already occupied by the GAGA factor and TFIID (TBP). The major change observed is the rapid targeting of HSF to the HSEs thanks to a nucleosome-free promoter. The mechanism by which binding of HSF trimers leads to a stimulation of transcription is not known. One suggestion is that HSF plays a role in accelerating the initiation of additional polymerase II molecules, which could be obligatorily coupled to the escape of the paused polymerase II (Lis and Wu, 1993). Figure 3.14 describes the activation process.

A paused RNA polymerase II is not restricted to *Drosophila* heat shock–gene promoters but is also featured in a variety of non-heat-shock-gene promoters. In *Drosophila*, these include the genes encoding α and β1 tubulin, polyubiquitin, and glyceraldehyde-3-phosphate dehydrogenase (GAPDH) (Lis and Wu, 1993). In this way, for example, the GAPDH enzyme can be induced faster if the cell shifts from an aerobic metabolic pathway to an anaerobic glycolytic respiratory pathway.

Returning to the aspect of HSF phosphorylation (discussed in Section 3.3.1.2), the phosphorylation step is required neither for the conversion to the trimeric state nor for the activation of DNA binding (Latchman, 1995; Morimoto et al., 1994a,b). On the contrary, evidence suggests that this phosphorylation step is related to transcriptional activation. A model is shown in Figure 3.15.

This regulatory mechanism can be affected by external stimuli, such as environmental stressors that activate protein kinases. Examples of such protein kinases are protein kinase A (PKA) of the cyclic adenosine monophosphate (cAMP) system and protein kinase C (PKC) of the PKC-dependent pathway and the calcium/calmodulin second messenger system (Hill and Treisman, 1995).

Hunter and Karin (1992) mentioned three levels of regulating transcription factor activity by phosphorylation: regulation of nuclear translocation,

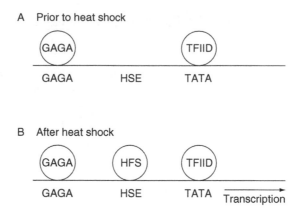

Figure 3.13 Proteins binding to the promoter of the hsp70 gene before (A) and after (B) heat shock. Abbreviations: GAGA = basal promoter binding element/protein; HSE = heat-shock element; HSF = heat-shock factor; TATA = basal promoter box; TFIID = TATA-box-binding protein complex. (After Latchman, D.S., *Eukaryotic Transcription Factors*, 2nd ed., Academic Press, San Diego, CA, 1995. With permission from Elsevier.)

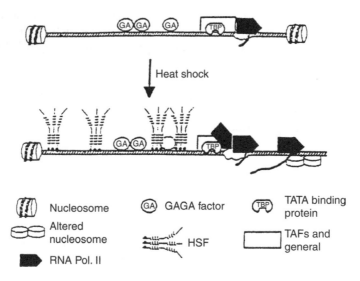

Figure 3.14 The architecture of the hsp70 gene promoter before and after heat shock. (After Lis, J.T. and Wu, C., *Cell* 74: 1–4, 1993. With permission.)

regulation of DNA binding, and regulation of transactivation. The phosphorylation shown in Figures 3.10 and 3.15 is a form of transactivation.

Transcription factors are sometimes composed of separate DNA-binding and transcriptional transactivation domains, which can be individually activated by phosphorylation or dephosphorylation. In most cases, phosphorylation of the transactivation domain results in a positive transactivation. Increased levels of PKA or PKC may contribute to this effect, probably in

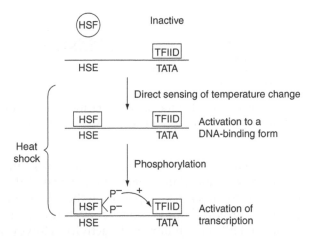

Figure 3.15 Stages in the activation of HSF in *Drosophila* cells. Abbreviations: HSE = heat-shock element; HSF = heat-shock factor; TATA = basal promoter box; TFIID = TATA-box-binding protein complex (After Latchman, D.S., *Eukaryotic Transcription Factors*, 2nd ed., Academic Press, San Diego, CA, 1995. With permission from Elsevier.)

combination with MAP kinases (Hill and Treisman, 1995; Hunter and Karin, 1992). An example of both positive and negative control of transcription factor activity is provided by the c-Jun transcription factor, a component of the activator protein-1 (AP-1) complex that binds to the (basal) promoter of numerous genes, illustrated in Figure 3.16.

Phosphorylation of the two sites near the N-terminal of c-Jun (Figure 3.16) increases transactivation (positive control), whereas phosphorylation of the three sites directly upstream of the DNA-binding domain inhibits DNA binding (negative control). Boyle et al. (1991) suggest that c-Jun is present in resting cells in this inhibited form, which can be released by site-specific dephosphorylation in response to, for example, PKC activation. Activation of PKC would, in addition to phosphorylating the transactivation domain, result in elevating the level of a protein phosphatase, dephosphorylating the sites near the DNA-binding domain and activating c-Jun DNA binding. This example illustrates the complexity of transcription factor activation by phosphorylation and the involvement of basal second messenger systems in transducing signals from environmental stressors; note that, for example, PKC can be activated by environmental stressors, such as phorbol esters and calcium ionophores.

In the beginning of Section 3.3, it is suggested that HSF in its monomeric form is stabilized by hsp70 (probably hsc70 in *Drosophila*) by transient interactions. Upon stress, the stress proteins bind to accumulated aberrant proteins and abolish their interactions with HSFs. The HSFs then trimerize and become activated. It should be mentioned here, however, that experiments conducted by Rabindran et al. (1994) demonstrated that, in general, the equilibrium between HSF monomers and trimers is not solely sensitive to changes in the concentration of (excess) HSP (cognates) and that some

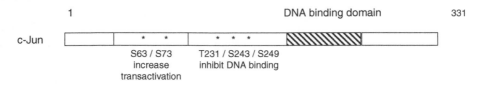

Figure 3.16 Schematic diagram of the transcription factor c-Jun, regulated by phosphorylation. (Adapted from Hunter, T. and Karin, M., *Cell* 70: 375–387, 1992. With permission.)

other mechanism for controlling trimerization and DNA-binding activity must exist. Jacquier-Sarlin and Polla (1996) demonstrated that the activation of HSF1 in human cell lines also depends on the redox state. Oxidants, such as hydrogen peroxide, may oxidize cysteine residues in the DNA-binding site of HSF and hamper its activity. However, hydrogen peroxide also induces the reductor thioredoxin and abolishes its inhibiting action. (The sensitivity of HSF for cellular redox changes is also discussed in relation to aging in Section 8.2.) Another aspect that seems to be involved in HSF activation is Ca^{2+} concentration. Calderwood and Stevenson (1993) reported that environmental stressors, such as heat shock, ethanol, and cadmium, activate the transmembrane enzyme PLC, resulting in transiently increased cytosolic Ca^{2+} levels (see Figure 2.1). They suggest that this Ca^{2+} ion, bound to calmodulin, is required for dissociation of the HSF/HSP complex, so that HSF can trimerize and HSP can bind to aberrant proteins.

Notwithstanding the above-mentioned findings, it remains a fact that the substrate level for hsp70 (hsc70) could be considered a translation of stress severity, since there is a correlation between these parameters. Hsp70 (hsc70) acts as the sensing mechanism between stress (accumulated proteins) and response (activation of HSF). According to Mosser et al. (1993), hsp90 (hsp83 in *Drosophila*) also interacts with HSF by binding to its inactive form, but it has not been shown to regulate its activity. Hsp90 (hsp83) could maintain HSF in the inactive form, whereas hsp70 (hsc70) stabilizes this complex by transient interactions (Mosser et al., 1993).

It has recently been found that human HSF assembles in analogous complexes containing most, if not all, of the proteins that were previously found to associate with steroid receptors (Voellmy, 1996). This finding is in line with an alternative proposal by Mosser et al. (1993) and resembles the configuration shown in Figure 3.4. The statement made by Hartl (1996) and mentioned in Section 3.2.1 that, under stressful conditions, hsp70 binds to denatured proteins in cooperation with hsp90 and smaller stress proteins supports the model shown in Figure 3.4. In the case of HSF binding and dissociation, it could mean that, under stressful conditions, the whole complex dissociates from HSF and binds to the denatured proteins.

Upon return from stress, HSF dissociates from the promoter of the gene it had activated. This process of dissociation and the involvement of hsp70 in this process, will be discussed in Section 3.3.3.

3.3.2 Posttranscriptional and translational regulation of stress proteins

Regulatory mechanisms acting at the levels of RNA processing, translation, and mRNA degradation are less well understood but have an equal, if not greater, effect on gene expression compared to transcriptional regulation (Yost et al., 1990). As mentioned in Chapter 1, interruption of splicing of precursor mRNAs is one of the effects observed in stressed cells. Further investigation has revealed that splicing was maintained during mild heat shock, whereas it was blocked at severe heat shock. This was observed with hsp83, one of the *Drosophila* stress proteins that contains an intron, and other proteins. To test whether this idea also holds for non-heat-shock-inducible gene products, the intron-containing alcohol dehydrogenase gene was placed under the control of an HSP70 promoter. At mild heat shock, the transcripts were accurately spliced. At severe heat shock, splicing stopped and intron-containing mRNA precursors accumulated (Yost et al., 1990). Summarizing these analyses, it may be concluded that interruption of splicing is exerted on all precursor mRNAs containing one or more introns but is dependent on the severity of the stress.

Immunological analysis of *Drosophila* embryos with antibodies to ribonucleoprotein (RNP) particles suggests that heat shock produces a change in the conformation of RNP complexes involved in splicing. It was suggested that hsp70 most likely plays a stabilizing role in functional spliceosomes and thus in protecting the splicing process to the extent possible (Yost et al., 1990).

Interruption of splicing may have acted as a selective pressure in the evolution of intronless HSPs in *Drosophila* species. Additional pressure may have been provided by the fact that precursor RNAs, blocked at splicing, can enter the cytoplasm and be translated to aberrant polypeptides. This, in turn, may have provided a driving force in evolution for decreasing the transcription of non-heat-shock genes during stress (Yost et al., 1990).

It is not yet clear why hsp83 and some other stress proteins, such as hsc71 and hsc72, still contain an intron or how they can partially circumvent the block in splicing. It is remarkable that the hsp83 intron is located directly upstream from the coding region and not somewhere in the open reading frame. The fact that the messages of these stress proteins contain one or more introns could help explain why the possibilities for processing these mRNAs under heat-shock conditions are limited. Under these conditions, hsp70 mRNA concentration is quickly built up. The half-life time under these conditions appears to be more than 4 hours. It was noticed, however, that during recovery from stress, mRNA concentration of hsp70 is quickly reduced, and after restoration of homeostasis, hsp70 level is low. To investigate the half-life time under normal unstressed conditions, hsp70 genes were given a metallothionein (MT) promoter, so that the gene could be expressed by copper without stress. It appeared that the half-life time of hsp70 mRNA is only 15 minutes under normal conditions.

Apparently, a mechanism for the degradation of hsp70 message exists in *Drosophila* cells (Yost et al., 1990). Jacquier-Sarlin et al. (1995) reported that hsp70 mRNA contains a 3'-untranslated AU-rich region with AUUUA motifs. This would be the element through which hsp70 and other stress proteins are regulated posttranscriptionally. Moreover, there is a cytoplasmic binding protein, called AU-binding factor, which forms complexes with a variety of labile RNA messages containing multiple reiterations of the pentamer AUUUA. Jacquier-Sarlin et al. (1995) further reported that stabilization of hsp70 mRNA by calcium ionophores and phorbol esters is probably mediated via PKC-activated phosphorylation of the AU-binding factor. According to the nature of the stimuli (e.g., heat shock or mitogen, i.e., the required velocity of the response), a given cell could use either genomic 5'-transcriptional regulatory elements or 3'-mRNA posttranscriptional elements, or both to finally control stress protein synthesis and provide an adequate response to the stress. Deactivation of the AU-binding factor after stress could lead to destabilization of the hsp70 mRNA, resulting in reduced half-life times.

The most dramatic effect of stress on the translational level is the change in protein synthesis. At a temperature elevation in *Drosophila* from 25 to 37°C, the translation of preexisting messages is stopped and preexisting polysomes disappear. As newly transcribed HSP mRNAs appear, polysomes are reformed on the HSP mRNAs and HSPs quickly become the major products of protein synthesis in the cell. The overall effect is not so much a reduced or increased translation efficiency, but rather a redirected protein synthesis (Yost et al., 1990).

The observation that preexisting mRNA and newly synthesized non-stress protein mRNA are not translated during heat shock, whereas HSP mRNAs are translated, has led to thorough biochemical studies. On the basis of these investigations, the following explanatory model has been developed: Preexisting (normal) 25°C mRNAs require a cap-binding factor for efficient translation, possibly to unwind secondary structures in the message leader. Heat shock inactivates this factor by dephosphorylation (Bensaude et al., 1996). HSP mRNAs, by virtue of their long unstructured leader (46% adenosine), are able to escape the requirement for cap binding and are therefore translated (Yost et al., 1990).

Is this just a consequence of stress to which the cell has not yet adapted, or is it a cellular action? One good reason for stopping "normal" protein synthesis is that there could be a shortage of free hsp70 (hsc70) chaperones for binding to nascent polypeptides, because the pool of HSPs (HSCs) is depleted owing to accumulation of denatured proteins from stress. The cell is then forced to give preference to synthesis of hsp's (Bensaude et al. 1996). Another reason may be a shortage of building blocks (amino acids) for protein synthesis and that preference has to be given to synthesis of stress proteins. An indication for this is that preference is given to the degradation of aberrant proteins over DNA processing and ribosome synthesis (see Section 3.2.4).

Some stressors, such as heat shock, calcium ionophores, and sodium arsenite, bring about a temporary or partial inhibition of translational initiation by phosphorylation of a eukaryotic initiation factor-2α (eIF-2α) for translation (Brostrom and Brostrom, 1998). It is not known, however, how the translation process of stress proteins can circumvent this inhibition.

Together, a restriction in the synthesis of normal proteins and, indirectly, a preferential synthesis of stress proteins is accomplished by factors such as des-ubiquitination of histones (see Section 3.2.4), a block in splicing, cap binding, and translational initiation. (The latter three aspects were discussed in this section.) The factors causing these interruptions or modifications are of a physical and chemical nature. A sharp differentiation between these factors is difficult to make, although, for example, a block in splicing may be predominantly caused by heat shock and not by chemical factors. Chemical factors probably have a greater effect on cap binding and translational initiation, in which phosphorylation/dephosphorylation steps and basal signal transduction systems are involved.

3.3.3   Regulation of recovery

When the stress is released, normal patterns of synthesis are restored. The restoration is a gradual two-phase process. First, after release of the stress, the cell continues to produce stress proteins exclusively for several hours, depending on the severity of the stress. Then, a gradual increase in normal protein synthesis and a gradual decline in stress protein synthesis is initiated. This second phase also lasts several hours. Preexisting messages are reactivated, while stress protein mRNAs are asynchronically repressed and degraded. Hsp70 is always repressed first, and hsp83 last (Yost et al., 1990). It is the reactivation of this degradation mechanism for stress protein mRNAs, together with the repression of the stress protein genes, that mediates repression of stress protein synthesis during recovery (Yost et al., 1990). The recovery process is also enhanced by partial restoration of solubility and activity of many aggregated proteins. In this renaturation process, at least hsp70 is involved (Bensaude et al., 1996).

For an efficient and complete recovery from the stress event, those conditions that had evolved to enable a stress response have to be removed. This involves restoration of normal protein synthesis including ubiquitination of histones and removal of the block in splicing, cap binding, and translational initiation and repression of stress protein synthesis including degradation of stress protein mRNAs and deactivation of stress protein gene expression. How all these conditions are fulfilled is not quite clear, but some comments on a few of them can be made.

The block in splicing is probably due to a change in conformation of RNPs, integrated in spliceosomes. On return to normal temperatures, this could be abolished if the HSPs have protected the RNP particles against irreversible damage. HSPs do protect the RNP particles to minimize the effect of the block in splicing and to promote a fast return to normal protein synthesis. This is not in conflict with the earlier observation that, if the cell has to choose between synthesis of stress proteins and normal proteins, preference is given to the former. This description only refers to heat stress. Whether chemical stressors cause the same effect and how this would be accomplished, is not clear.

With regard to degradation of mRNAs for stress protein, Section 3.3.2 discusses stabilization of an AU-binding factor mediated by PKC. Deactivation of this factor is believed to lead to destabilization of stress protein mRNAs and reduced translation.

Yost et al. (1990) suggest that hsp70 plays an important role in the recovery process and particularly in the degradation of aberrant proteins (discussed below) and the deactivation of stress protein gene expression. It was noticed that for a given heat treatment, a specific quantity of HSP is always produced before normal protein synthesis at 25°C in *Drosophila* is restored (Shi et al., 1998; Yost et al., 1990). However, at least in *Drosophila*, hsp70 is only necessary during stress and is undesirable without stress. During normal conditions, its role is fulfilled by its cognate, while hsp70 concentration in *Drosophila* is then hardly detectable. The fact that, after heat shock, processing of hsp70 continues for several hours does not contradict this. HSFs bound to HSP genes have first to dissociate from the promoter, and, according to step 6 in Figure 3.10, HSPs are involved in this process. Maybe first, extra (free) HSP is produced after heat shock to stop its own processing by assisting in the dissociation of HSFs. Shi et al. (1998) showed that hsp70 interacts directly with the transactivation domain of HSF1 in human cells and represses heat-shock gene transcription. As neither the activation of HSF1 DNA-binding nor inducible phosporylation of HSF1 was affected, the primary autoregulatory role of hsp70 is to negatively regulate HSF1 transcriptional activity. This negative regulation has also been demonstrated in experiments by Baler et al. (1996). An indirect proof of the involvement of HSPs in the down regulation of HSP gene transcription is the incorporation of amino acid analogs in HSP synthesis. The newly synthesized chaperone is nonfunctional and cannot repress HSF activity. Consequently, HSF, which is a stable protein and not dependent on continuous protein synthesis, remains in a transcriptionally active state. Thus, inducible transcriptional activity of HSF is intimately linked to the autoregulatory aspects of hsp70 and not with its constitutively present cognate, hsc70 (Shi et al., 1998).

This form of autoregulation can also be viewed in a broader context, that is, the return to homeostasis: Referring to the model for transcriptional regulation (Figure 3.7), the trigger for the response to stress would be the appearance of aberrant or denatured proteins that would compete with HSF for (transient) association with HSP (HSC). This model could also provide an explanation for the down regulation of the stress protein synthesis after stress response; that is, as a result of stress response the pool of free stress proteins increases and the HSF/HSP equilibrium shifts in the direction of association, or in other words, HSFs appear in the monomeric, inactivated form. As a consequence, transcription and translation of stress proteins is attenuated (Morimoto et al., 1990b, 1994b). Mosser et al. (1993) have tested the role of human hsp70 in the negative control leading to recovery of homeostasis, and on the basis of these tests they developed the model shown in Figure 3.17.

hsp70 is used to destabilize the interaction between trimeric HSF and HSE. Then hsp70 or its cognate stabilizes the monomeric inactive form. This negative regulation is relieved when substrate levels for hsp70/hsc70 (denatured or aberrant proteins) are elevated as a result of cellular stress.

As described in Section 3.2.4, ubiquitin is involved in the degradation of unfolded or wrongly folded proteins (see Figure 3.6). Ubiquitin tags proteins destined for degradation, but hsp70 also plays a role in degradation, as discussed above. Schlesinger (1990) has combined these two activities (of hsp70 and ubiquitin) in a competitive model, illustrated in Figure 3.18.

In his model (Figure 3.18), Schlesinger (1990) distinguishes a "rescue" pathway, in which hsp70 is involved, in competition with a "degradation" pathway, in which (poly)ubiquitin is involved. What cannot be repaired will be degraded, so that the building blocks for protein synthesis can be reused, and a return to normal conditions is stimulated. In *Drosophila*, after return to homeostasis, excess hsp70 proteins are irreversibly inactivated by sequestration in granules (Feder et al., 1992). This aspect will be discussed further in Chapter 7.

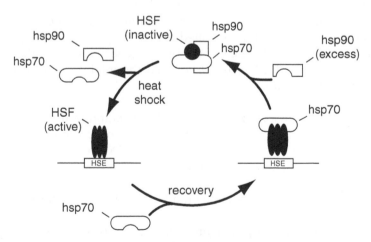

Figure 3.17 Model of regulation of HSF activity by hsp70. During recovery, hsp70 binds to HSF in the trimeric form, destabilizes it, and refolds the monomer. Thus, hsp70 stabilizes HSF in the inactive form, together with, or assisted by, hsp90. Abbreviations: HSF = heat-shock factor; hsp = heat-shock protein. (After Mosser, D.D. et al., *Mol. Cell. Biol.* 13: 5427–5438, 1993. With permission from the American Society for Microbiology.)

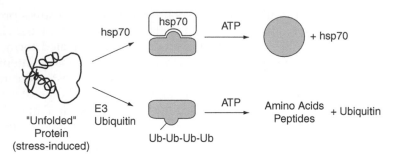

Figure 3.18 A competitive degradation model with a "rescue pathway" (hsp70) and a "degradation pathway" (ubiquitin). Abbreviations: ATP = adenosine triphosphate; hsp = heat-shock protein; Ub = ubiquitin. (After Schlesinger, M.J., in *Stress Proteins: Induction and Function*, Schlesinger, M.J. et al., Eds., Springer-Verlag, Heidelberg, 1990. With permission.)

3.4 Multiple pathways for stress protein gene expression

Figure 3.7 shows a general scheme of how stressors initiate stress protein gene expression and induce stress proteins, which contribute to the cellular stress response. This section focuses on answering questions such as:

What are the intracellular signals activating HSF?
Are there specific regulating pathways in response to different signals?

Watowich and Morimoto (1988) studied the expression of human hsp70 and grp78 genes in response to several inhibitors of cellular metabolism. They found that all inhibitors activated grp78, the gene expressing the main stress protein for the ER, whereas the hsp70 gene, expressing the constitutively present and inducible hsp70, divided the inhibitors in three classes, according to its response:

1. No effect on hsp70: inhibitors of glycosylation in the ER
2. Activation of hsp70: amino acid analogs and cadmium
3. Repression of hsp70: 2-deoxyglucose (2-DG, a glucose analog) and A23187 (a calcium ionophore).

The following comments can be made regarding these three items:

1. Inhibitors of glycosylation only affect grp78 (BiP), since glycosylation is an ER-related process. It leads to accumulation of incompletely assembled proteins in the ER, which probably induces the expression of grp78 via a specific transcription factor.
2. Amino acid analogs affect primary structures and proper folding, leading to accumulation of aberrant proteins in both the cytoplasm and the ER and consequently to the expression of both the hsp70

gene and the grp78 gene. Cadmium affects proteins containing sulfhydryl groups, leading to accumulation of aberrant proteins in the cytoplasm and expression of the hsp70 gene. Cadmium also increases intracellular Ca^{2+} concentration, resulting in the expression of the grp78 gene (Brostrom and Brostrom, 1998) (discussed in Chapter 5). Moreover, there is a direct pathway for cadmium to induce MT expression. This will be discussed below.

3. A23187 is a calcium ionophore and increases transiently cytosolic Ca^{2+} levels, disturbing basal signaling pathways in which Ca^{2+}/calmodulin is involved. Possible disturbance of Ca^{2+} homeostasis in the ER by calcium ionophores would impair the glycosylation process and lead to the activation of the grp78 gene (Li and Lee, 1991). 2-DG, a glucose analog, could stimulate glucose deprivation and grp78 gene expression. Moreover, both chemicals appeared to reduce hsp70 mRNA levels posttranscriptionally. The same effect was also reported by Sanders (1990) and may explain the observation in Section 3.2.1.3 that when GRPs are induced by glucose or oxygen deprivation, other stress proteins seem to be repressed. However, Jacquier-Sarlin et al. (1995) reported seemingly opposite effects by calcium ionophores and the phorbol ester PMA. These compounds would stabilize hsp70 and hsp90 mRNA via PKC-mediated phosphorylation. Glycosylation is believed to be affected by oxygen deprivation because oxygen deprivation causes a lower production of guanosine triphosphate (GTP) because of reduced activity of the citric acid cycle. GTP is required for transport of glucose to the ER.

Watowich and Morimoto (1988) also tested the response on heat shock: hsp70 gene expression was induced by heat shock, whereas grp78 gene expression was only slightly induced by the same heat shock and in a later stage. They suggest that this is due to the HSF for inducing the grp78 gene not being constitutively present and having to be induced and synthesized first. It may be concluded that the HSFs for activating the hsp70 and grp78 genes should be different. This conclusion is of interest from a control point of view and will be analyzed here in more detail.

Signal transduction from organelles to the nucleus is more complicated than signal transduction from the cytoplasm to the nucleus, because an extra intracellular membrane has to be crossed. Cox et al. (1993) reported on the existence of a special pathway in *Sacharomyces cerevisiae* for expressing BiP (identical to grp78) for the ER. The pathway is called "unfolded protein response" (UPR). What matters here is that the promoter of the KAR2 gene, expressing BiP, contains a special UPR element (UPRE). This UPRE is bound by a special transcription factor, the unfolded protein response (transcription) factor (UPRF) (Mori et al., 1993). If these findings also apply to higher eukaryotes, and that seems to be the case (Ivessa, 2002; Patil and Walter, 2001), they support the presence of a special transcription factor for the gene expressing the stress protein for the ER. They also suggest the existence of two signaling pathways for the expression of this stress protein: one for signals from the cytoplasm, related to disturbances in cytoplasmic processes, the other for disturbances of processes in the ER, such as disturbance of glycosylation. The UPR pathway is illustrated in Figure 3.19.

Kohno et al. (1993) reported on the presence of an HSE in the KAR2 promoter, upstream from the UPRE. Both elements are independent, and binding by their respective transcription factors (HSF and UPRF) gives an additive effect, which implies different mechanisms. Kohno et al. (1993) further reported that the UPRE is also conserved in the promoter of the mammalian grp78 gene, but that a similar HSE was not identified in this homolog. This could explain why Watowich and Morimoto (1988) found that mammalian grp78 gene expression was only slightly induced by heat shock. If the promotor of the hsc72 gene expressing the stress protein for the ER of *Drosophila* does not possess an HSE either, it could also explain the observation by Rubin et al. (1993) that the hsc72 protein is not inducible by heat.

In Figure 3.19, it is assumed that the UPRF is constitutively present, whereas Watowich and Morimoto (1988) assumed that the HSF for expression of the grp78 gene has to be induced (see above). That seems to be contradictory. Watowich and Morimoto (1988) also suggest that an increase in improperly folded or denatured proteins will induce hsp70 and grp78, with the specificity of the induction being dependent on whether the target of damage is confined to the ER. This suggests a special pathway system, like the model in Figure 3.19. Such a special pathway system could provide another explanation for why glucose or oxygen deprivation cause the induction of GRPs and the repression of other stress proteins as mentioned above. Watowich and Morimoto (1988) suggest that simultaneous repression of hsp70 is posttranscriptionally regulated.

Another reason for the different response to some stressors by hsp70 and grp78 genes, as reported by Watowich and Morimoto (1988), could be the difference in redox state between the cytoplasm and the ER: The first has a reducing environment, the second an oxidizing environment. This could lead to different effects in the compartments. Apart from physical damage at high temperatures, heat shock resembles an oxidizing agent, for example, by causing "electron leakage" and subsequent production of reactive oxygen species (ROS) (discussed in Chapter 4). The consequence is that damage by heat shock (concentration of denatured or aberrant proteins), depending on its severity, can be different in the cytoplasm than in the ER (Brostrom and Brostrom, 1998; Huang et al., 1994).

What applies to heat shock is also valid in general for different effects of reducing and oxidizing agents: Reducing agents exert their effects initially in the oxidizing environment of the ER, oxidizing agents in the reducing environment of the cytoplasm. For instance, dithiothreitol (DTT), a thiol-reducing agent, induces grp78 (Brostrom and Brostrom, 1998) and even suppresses the induction of hsp70 by heat shock, presumably owing to prevention of heat-induced protein oxidation by DTT (Huang et al., 1994).

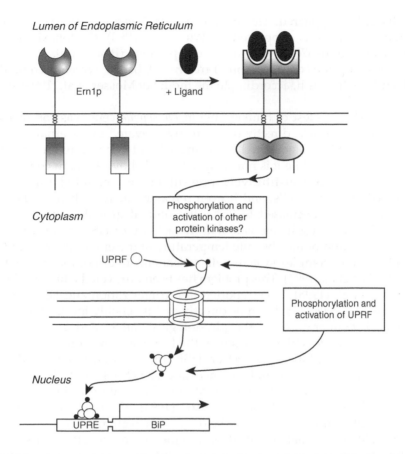

Figure 3.19 A model for a possible pathway of signaling from the ER to the nucleus. According to this model, the accumulation of unfolded proteins in the ER generates a ligand that binds to the transmembrane protein, activating a kinase domain at the cytosolic side. The signal is then transduced by phosphorylation of a UPRF, leading to transcriptional activation of the BiP gene. Abbreviations: Ern1p = transmembrane protein in the ER; UPRF = unfolded protein response (transcription) factor; UPRE = unfolded protein response element; BiP = binding protein. (After Mori, K. et al., *Cell* 74: 743–756, 1993. With permission.)

The observation that the HSF that induces hsp70 gene expression is constitutively present was tested and reported by Morimoto and Milarski (1990): Cells, heat shocked in the presence of cycloheximide, an inhibitor of protein synthesis, by inhibiting translation elongation, transiently activate HSF and hsp70 gene expression for only 1 h. Cycloheximide then blocks the appearance of HSF. According to Huang et al. (1994), cycloheximide requires an incubation time of about 1 h to become effective. In contrast, cells treated with cadmium, in the presence of cycloheximide, maintain high levels of HSF for up to 6 h. Upon addition of cadmium to the first group of cells, HSF reappeared. This proved that HSF was still present but inactivated. This view is supported by Wu et al. (1990). Cycloheximide also liberates hsc70/hsp70 bound to nascent proteins (Baler et al., 1996). This could build up the concentration of free hsc70/hsp70, which then binds to HSF and inactivates it, according to the model of Mosser et al. (1993) shown in Figure 3.17.

A transfection test of human HSF in *Drosophila* cells was conducted to determine if HSF was directly or indirectly activated by external stressors. The human HSF was activated at the induction temperature for *Drosophila* (more than 30°C), while it is normally activated above 40°C. In contrast, the *Drosophila* HSF is constitutively present in the activated form, when expressed in human cells, and it even remained active when the host cell temperature was decreased to 25°C, the normal growth temperature for *Drosophila*. These results indicate that the activity of HSF *in vivo* cannot be a simple function of the absolute temperature of the environment, but there must be a more complex system with sensors upstream of the HSF (Clos et al., 1993; Wu et al., 1994). This problem has been unraveled a little bit further by Zhong et al. (1998). They reported that trimerization and DNA binding of purified *Drosophila* HSF can be directly and reversibly induced *in vitro* by heat-shock temperatures in the physiological range, and by an oxidant, hydrogen peroxide. Other inducers of the heat-shock response, such as ethanol and arsenite, have no effect on HSF trimerization *in vitro*, indicating that these inducers act by indirect mechanisms (Zhong et al., 1998). Section 3.3.1.3 discusses the involvement of the Ca^{2+}/calmodulin complex as described by Calderwood and Stevenson (1993). The indirect induction by arsenite will be discussed further below and in Section 4.2.3.

It could be speculated that, at least in the case of elevated temperatures, conformational changes in the monomeric HSF and exposure of hydrophobic sequences leading to trimerization are a causal factor in directly activating HSF. This observation of direct activation of the *Drosophila* HSF by heat shock does not impede the action of other factors involved in HSF activation, such as dissociation of HSPs from HSF by dissociation factors and binding of dissociated HSPs to denatured proteins. Zhong et al. (1998) also realize that secondary factors influence HSF activity in *Drosophila* cells.

Morimoto and Milarski (1990) reported on a "pathway analysis": When three types of genes in human cells are compared (an HSP gene, e.g., hsp70; a metallothionein gene, e.g., MT I/II; and a glucose-regulated gene, e.g., grp78), the following observations are made:

1. The hsp70 and grp78 genes are activated by heat shock.
2. All three gene types are activated by cadmium.
3. The grp78 gene is activated by glucose deprivation.

Morimoto and Milarski (1990) suggest that there are multiple pathways for activating the transcription factor, with one common factor upstream of it. In another paper (Morimoto et al., 1990a), a more differentiated approach is followed, and it is suggested that there are metal-ion sensitive intermediates in the pathway for HSF activation. The following comments can be made about this:

1. Heat shock may physically cause denaturation of proteins in all compartments and membranes of the cell, depending on the temperature applied and the duration of the heat shock. For instance, Huang et al. (1994) exposed human HeLa cells to temperatures of 42°C and 45°C, both in the presence of DTT, a suppressor of heat shock. At 45°C, HSF activity could no longer be suppressed, resulting in heat-shock response. If denatured proteins finally accumulate in all compartments, both hsp70 and grp78 are induced, but along different transcription pathways because the grp78 gene does not contain an HSE. From this point of view, these results need not be in conflict with the findings of Watowich and Morimoto (1988) described above.

2. MT genes, from yeast to man, contain metal-responsive elements (MREs) in their promoters, and there exist, at least in mammalian species, several metal-responsive transcription factors (MRTFs) binding MREs, such as the metal-responsive transcription factor (MTF) (Koropatnick and Leibbrandt, 1995). Thus, cadmium is able to activate MT genes directly. For further comments on activation of stress protein genes by cadmium, see the tests of Watowich and Morimoto (1988) described above.

All these considerations have led to the development of a speculative model, shown in Figure 3.20.

The model in Figure 3.20 complies with the studies of Watowich and Morimoto (1988) and Morimoto and Milarski (1990) and is based on their multiple pathway concept. The heart of the general pathway for inducing stress proteins is formed by the combination HSF/HSE, activated by the accumulation of aberrant proteins in the cytoplasm. The stressors causing this accumulation are represented in Figure 3.20 by "heat shock" (No. 1). However, heat shock may also affect HSF in a direct way. Heat shock is considered a general stressor because it also exerts its effects in organelles, inducing stress proteins via other routes in addition to HSF/HSE. There are other stressors, however, that exert their effects in organelles only. The signaling pathway from the ER to the nucleus is used to illustrate this. Glucose deprivation (No. 2) represents the type of stressor activating this pathway, by means of a UPRF/UPRE combination. A third group is formed by heavy metals, which induce MT directly and stress proteins indirectly. This pathway, containing an MTF/MRE combination, is represented by cadmium (No. 3).

It is worth noting that the promoter of the CUP1 gene of the yeast *S. cerevisiae*, the gene expressing MT, appears to contain an HSE. At heat shock and glucose limitation, the CUP1 gene is induced directly and additional MT is synthesized (Sewell et al., 1995). Sequences resembling HSEs in

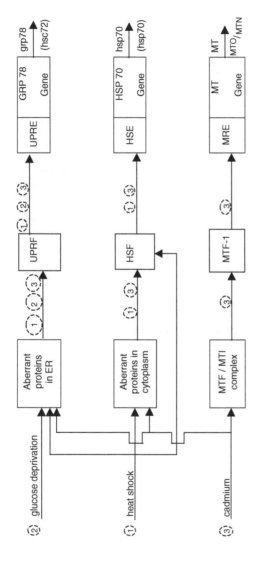

Figure 3.20 Speculative model for the induction of stress proteins and metallothionein (MT) in response to glucose deprivation, heat shock, or cadmium. Glucose deprivation only induces grp78, heat shock induces both hsp70 and grp78, whereas cadmium induces MT directly, and both hsp70 and grp78 indirectly. Abbreviations: UPRF = unfolded protein response element; ER = endoplasmic reticulum; HSE = heat-shock element; HSF = heat-shock factor; MTF = metal-responsive transcription factor; MTI = metallothionein transcription inhibitor; MRE = metal-responsive element; MT = metallothionein; MTO/MTN = specific MTs in *Drosophila*; grp = glucose regulated protein; hsp = heat-shock protein; hsc = heat-shock cognate. *Drosophila* names of end products are in parentheses. *Note:* Some abbreviations refer to proteins in human cells. It is supposed that identical proteins or homologs exist in *Drosophila* cells.

promoter sequences of *D. melanogaster* and human MT genes were not detected (Silar et al., 1991). Probably, the same applies to sequences resembling MRE promoter sequences in *Drosophila* HSP genes. This would be a possible way for heavy metals to induce transcription of stress protein genes directly. Therefore, these observations were not considered when developing the scheme in Figure 3.20.

From Figure 3.20, it may be concluded that whether a stress protein gene is eventually activated or not is determined by the presence of specific receptor sequences, such as HSE and UPRE, in the promoter of the stress protein gene. The affinity for the corresponding transcription factor may vary considerably, resulting in a different activation of transcription (see Section 3.3). The multiple pathway system also explains why glucose deprivation causes the induction only of GRPs (represented by grp78 in Figure 3.20) and not that of other stress proteins, such as hsp70.

The question remains concerning to what extent basal signaling systems, such as second messenger systems and the MAPK system, are involved in transducing stress signals and in activating transcription of stress protein genes. The MAPK system contains special kinases that are activated by heat shock and contribute to the activation of stress protein genes. It has been reported that existing SHSPs are phosphorylated and activated by MAPK upon stress. The second messenger system with the common pathway of protein kinase C (PKC) and calcium/calmodulin is also involved in transducing proteotoxic stress signals. For instance, the calcium ionophore A23187 causes release of sequestered calcium ions in the ER, leading to transiently increased Ca^{2+} concentrations in the cytoplasm. Bound to calmodulin, it mediates PKC activity and subsequent activation of transcription factors. It also exerts posttranscriptional effects by inhibiting translational initiation (Brostrom and Brostrom, 1998). Another example of Ca^{2+} involvement was provided by Cornelius (1996), who demonstrated that heat-shocked *Drosophila* cells showed a sustained increase in cytosolic Ca^{2+} level, as well as an increased concentration of inositol triphosphate (IP_3), an intermediate product of the second messenger system releasing Ca^{2+} from the ER. Notwithstanding the increased IP_3 concentration, Cornelius (1996) took into account that the increased Ca^{2+} concentration could have arisen from extracellular sources owing to plasma membrane damage by heat shock. Chemical stressors inducing HSPs, such as arsenite, cadmium, and ethanol, also increased IP_3 levels, but a sustained increase in Ca^{2+} could not be demonstrated. Cornelius (1996) therefore concluded that Ca^{2+} is not a common factor in activating HSF in response to proteotoxic stress. This is contradictory to what Calderwood and Stevenson (1993) report; that is, that Ca^{2+} is required for dissociation of the HSF/HSP complex and activation of HSF (see also Section 3.3.1.3). It also seems to be different for chemicals causing oxidative stress but inducing HSPs. This will be discussed further in Chapter 4.

However, Munks and Turner (1994) reported an inhibiting effect of ethanol on HSP synthesis in cultured *Drosophila* cells. However, in these cells the alcohol dehydrogenase enzyme, converting alcohols to aldehydes, was not detectable. This demonstrates how difficult it is to draw general conclusions on the basis of *in vitro* tests. It also shows that stress effects and responses are cell specific and that the means of attack is not uniform. The most striking example is presented by arsenite, the active trivalent form of the transition metal As. Arsenite causes proteotoxic, oxidative, metal, and genotoxic stress, and elicits, accordingly, several cellular responses. Restricting it to the subject of this chapter, arsenite causes proteotoxicity by binding to sulfhydryl groups of proteins (Sok et al., 2001). The induction of HSPs is caused not only by the accumulation of aberrant proteins, but also by the activation of MAPKs by arsenite (Cavigelli et al., 1996; see Section 4.2.2). Thus, the involvement of basal signal transduction systems in inducing HSP synthesis indirectly is clear.

In summary, environmental stressors may damage cellular proteins, both structural proteins in filaments and membranes and enzymatic proteins essential in cellular processes. Accumulation of these aberrant proteins leads to induction of stress proteins, which participate in the stress response and contribute to a return to homeostasis. This is mainly accomplished by:

1. Stabilizing essential proteins of structures, especially around the nucleus, and of organelles, such as ribosomes and spliceosomes
2. Participating in the refolding of denatured proteins
3. Contributing to the degradation process of aberrant proteins

Many aspects of these processes have not yet been resolved, and particularly, the regulation of gene expression and recovery still evokes many questions. It is clear, however, that the various groups of stress proteins play a central role in protecting the cell against proteotoxic stress and in repairing damage to proteins and enzymes.

chapter 4

The oxidative stress response system

4.1   Introduction

Oxidative stress is part of the more general term *environmental stress*, but the former primarily causes the response of components of the oxidative stress response system, whereas the latter also includes subjects such as proteotoxic, metal, and genotoxic stress.

Oxidative stressors may cause the generation of reactive oxygen species (ROS), including free radicals such as the superoxide anion ($O_2^{\bullet-}$), the hydroxyl ($OH^{\bullet}$) and the nitric oxide ($NO^{\bullet}$) radical, as well as the nonradical intermediates, such as hydrogen peroxide (H_2O_2) and singlet oxygen ($^1O_2$). Oxidative stressors are intermediates of oxygen reduction processes, such as respiration, of redox reactions, and of the metabolism of chemicals. During normal cellular respiration, ROS are constantly produced at a low rate in both eukaryotes and prokaryotes. At these low concentrations, ROS can stimulate cell proliferation by acting as second messengers (Storz and Polla, 1996).

Upon exposure to environmental stressors, extra amounts of ROS may be generated that can accumulate to toxic levels. At these elevated concentrations, ROS lead to oxidative stress resulting in cytotoxicity and damage of cellular structures, such as membranes (Jamieson and Storz, 1997; Storz and Polla, 1996). These effects may initiate a series of events that culminate in apoptosis (Flores and McCord, 1997; Fuchs et al., 1997) or necrosis (Richter and Schweizer, 1997).

The discussion above may make it clear that ROS are both essential and harmful for aerobic cells (Chaudère, 1994). During evolution, natural selection formed the basis for developing a defense system against ROS, to maintain the balance between essential levels and harmful effects. This defense system can be divided into two major groups:

1. Enzymatic systems to maintain harmless levels of activated oxygen species by reducing excess to H_2O
2. Nonenzymatic scavengers to remove free radicals formed during these reduction processes or during deleterious reactions between excess ROS and cellular macromolecules

In the following sections, the sources of ROS, their effects, the response systems, and their integrated reactions will be discussed in detail. This will be done in two parts: In the first part, the enzymatic reduction processes and nonenzymatic scavenging systems in the homeostatic cell will be reviewed, and in the second part the events, effects, and responses will be discussed in relation to oxidative stress. However, a distinct separation of these aspects cannot be made between the two sections (4.2.1 and 4.2.2). The general division between these aspects under homeostatic and stressful conditions should be considered as a framework for describing this complicated subject.

4.2 Cellular response to reactive oxygen species

4.2.1 Generation of reactive oxygen species and cellular response under homeostatic conditions

Under homeostatic conditions, the superoxide anion radical ($O_2^{\bullet-}$) is predominantly produced in the respiratory chain of mitochondria by autoxidation of reduced components in the chain (Winyard et al., 1994). Although this production of $O_2^{\bullet-}$ may be considerable (up to 1 to 2% of O_2 reduction to H_2O, depending on the oxygen tension), the concentrations of $O_2^{\bullet-}$ remain very low (in the picomolar range). This is because $O_2^{\bullet-}$ is rapidly dismutated to hydrogen peroxide ($2O_2^{\bullet-} + 2H^+ \rightarrow H_2O_2 + O_2$) by superoxide dismutase (SOD) (Richter and Schweizer, 1997; Wolin and Mohazzab-H, 1997). Another important source for generating $O_2^{\bullet-}$ may be due to redox cycling, at which $O_2^{\bullet-}$ is produced during the regeneration step (discussed further in the next section). $O_2^{\bullet-}$ is also generated in the xanthine dehydrogenase pathway. In this pathway, adenosine monophosphate (AMP) is degraded in the cytosol to xanthine and subsequently to uric acid. The latter step is catalyzed by xanthine dehydrogenase (xanthine + H_2O + $NAD^+ \rightarrow$ uric acid + NADH + H^+). However, about 10% of the enzyme is present as an oxidase, which transfers electrons to O_2 to form $O_2^{\bullet-}$ (xanthine + H_2O + $2O_2 \rightarrow$ uric acid + $2O_2^{\bullet-} + 2H^+$) (Winyard et al., 1994; Wolin and Mohazzab-H, 1997).

If $O_2^{\bullet-}$ concentrations cannot be maintained at the required level by dismutation to H_2O_2 or by scavenging (discussed below), then excess $O_2^{\bullet-}$ may have damaging effects. First, $O_2^{\bullet-}$, although not very reactive, may inhibit enzyme activity by penetrating the active sites of the enzymes, thereby reducing the metal responsible for the enzyme's activity (Chaudère, 1994; Flores and McCord, 1997). Targets for inhibition by $O_2^{\bullet-}$ include nicotinamide adenine dinucleotide hydride (NADH) dehydrogenase, ATPase, and Krebs cycle enzymes (Richter and Schweizer, 1997). Inhibition of these groups of enzymes may have an effect on respiration, adenosine triphosphate (ATP) synthesis, and the structure of metabolic pathways (discussed further together with effects owing to changes in cellular redox state). Second, in acidic microenvironments of cell membranes, $O_2^{•-}$ may be protonated into the perhydroxyl radical $HO_2^•$. This radical is more reactive than $O_2^{•-}$ and is cytotoxic (Chaudère, 1994; Rice-Evans, 1994). Third, in the presence of the nitric oxide radical (NO$^•$), superoxide anion radical may form peroxynitrite (ONOO$^-$), which decomposes into hydroxyl radicals in an iron-independent reaction (Koster and Sluiter, 1994). ONOO$^-$ is an effective oxidant of thiol groups of enzymes by means of nitrosation, forming $RSNO_x$ (Richter and Schweizer, 1997; Wolin and Mohazzab-H, 1997). An example is the critical thiol group of glyceraldehyde-3-P-dehydrogenase (GAPDH), an important enzyme of glycolysis (Wolin and Mohazzab-H, 1997).

The enzyme superoxide dismutase fulfills an important role in converting $O_2^{•-}$ into H_2O_2. The intracellular SOD activity is concentrated in the mitochondrium as MnSOD and in the cytosolic compartment as Cu/ZnSOD. The activity of SOD is regulated through biosynthesis and stimulated by increased oxygen tension or chemical compounds (Yu, 1994). For instance, during cell cycling in *Drosophila*, MnSOD is regulated by mitogen-activated protein kinase (MAPK) signaling in response to an increased flux of superoxide anion radicals generated from enhanced respiratory demand (Duttaroy et al., 1997). The MAPK signal probably activates the MnSOD gene via the transcription factor AP-1, which subsequently may bind to one or more of the special response elements present on the promoter of the MnSOD gene, such as metal responsive element (MRE), antioxidant responsive element (ARE) and xenobiotic responsive element (XRE) (Duttaroy et al., 1997). The ARE is involved in oxidative stress response and is equivalent to the electrophile response element (EpRE) (Nebert et al., 2000). The various types of responsive elements in the MnSOD promoter illustrate the central position of this enzyme in the defense against environmental stress.

However, there is another important source of H_2O_2; that is, via a two-electron transfer, oxygen can enzymatically be reduced to H_2O_2 by, for example, monoamine oxidase (MAO). This enzyme is a flavoprotein ubiquitously expressed in higher eukaryotic organisms and localized on the outer mitochondrial membrane. It catalyzes an oxidative deamination, at which O_2 is converted to H_2O_2, whereas the substrate is deaminated and oxidized (Hauptmann and Cadenas, 1997; Kaul and Forman, 2000).

Generally, H_2O_2 is not reactive enough to oxidize many organic molecules in an aqueous environment. Nevertheless, it is a biologically important oxidant, because of its ability to generate the very reactive hydroxyl radical (OH$^•$). The biological importance of H_2O_2 also arises from its ability to diffuse through hydrophobic membranes with relative ease, as shown in mitochondrial H_2O_2 leakage. The basis for this resides in its nonionized state (Yu, 1994).

Hydrogen peroxide is normally decomposed to H_2O through two enzyme systems: catalase (CAT) and glutathione peroxidase (GSH-P$_x$). Catalase is a major antioxidant defense enzyme, predominantly present in peroxisomes of the cell, and it contains iron in its active site (Diplock, 1994). Catalase is also expressed in the cytosol of *Drosophila melanogaster* (Mockett et al., 2003). It catalyzes the reduction to H_2O ($2H_2O_2 \rightarrow 2H_2O + O_2$). The two electrons required to regenerate the oxidized iron are provided by nicotinamide adenine dinucleotide phosphate hydride (NADPH) (Diplock, 1994; Hauptmann and Cadenas, 1997; Yu, 1994). Most species exhibit GSH-P$_x$, of which the hydrophilic isoform is located in the cytosol and the mitochondrial matrix, whereas the hydrophobic isoform is membrane-bound (Diplock, 1994). GSH-P$_x$ catalyzes the reduction of H_2O_2 and the reduction of organic hydroperoxides (ROOHs). It participates in the reaction by using glutathione (GSH) as a substrate ($H_2O_2 + 2GSH \rightarrow GSSG + 2H_2O$). The disulfide (GSSG) is regenerated by glutathione reductase, using NADPH as an electron donor ($GSSG + NADPH + H^+ \rightarrow 2GSH + NADP^+$). CAT and GSH-P$_x$ have different substrate affinities. In the presence of low H_2O_2 levels, ROOHs are preferentially reduced by GSH-P$_x$; however, at high H_2O_2 concentrations, they are metabolized by CAT (Diplock, 1994; Hauptmann and Cadenas, 1997). According to Beckmann and Ames (1997), there are differences between organisms, for example, GSH-P$_x$ plays an important role in mammals but is absent in nematodes and insects, including *Drosophila* (Orr and Sohal, 1994). The decomposition of H_2O_2 in flies is fulfilled by CAT (Sohal et al., 1995). In the case of *D. melanogaster*, cytosolic CAT probably fulfills this role.

Hydrogen peroxide is also subjected to one-electron reduction with formation of the potent electrophilic hydroxyl radical (OH$^\bullet$). This is a Fenton-driven reaction, by which reduced transition metal complexes, generally involving Fe^{2+} or Cu^+, facilitate the decomposition of H_2O_2 ($H_2O_2 + Fe^{2+} \rightarrow OH^- + OH^\bullet + Fe^{3+}$). It can exert different types of biological damage, for example, site-specific (DNA) or in the bulk solution (nonspecific mechanism), but always in the neighborhood of its site of formation owing to its very short half-life.

The moderate chemical reactivity of H_2O_2 is substantially enhanced by this side reaction (Hauptmann and Cadenas, 1997). It was originally suggested that Fe^{3+}/Cu^{2+} was regenerated by the superoxide anion radical ($O_2^{\bullet-}$ + $Fe^{3+}/Cu^{2+} \rightarrow Fe^{2+}/Cu^+ + O_2$), but because $O_2^{\bullet-}$ concentrations are in the picomolar range, this is now considered to be unlikely. Glutathione and ascorbate (vitamin C) are considered better candidates for reducing the transition metals (Chaudère, 1994). Considering that, under normal physiological circumstances at neutral pH, iron is ligated, it is now believed that a ligated-iron ion of a higher oxidation state, a ferryl ion (FeO^{2+}), participates in the decomposition of H_2O_2 (Chaudère, 1994; Koppenol, 1994).

Figure 4.1 summarizes the enzyme systems participating in the cellular defense against oxidative stress and contributing to a cellular homeostasis by metabolizing endogenous harmful intermediates. The hydroxyl radical is a major component of oxidative damage and is considered the most potent

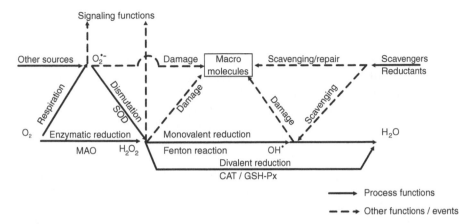

Figure 4.1 Sources of H_2O_2 and its metabolism during cellular homeostasis. If the steady state concentrations of $O_2^{\bullet-}$ and H_2O_2 cannot be maintained, then signaling functions are disturbed. Damage will arise from excess production of $O_2^{\bullet-}$, H_2O_2, and OH•. As a consequence, the cell will no longer be in homeostasis. Abbreviations: MAO = monoamine oxidase; $O_2^{\bullet-}$ = superoxide anion radical; SOD = superoxide dismutase; H_2O_2 = hydrogen peroxide; OH• = hydroxyl radical; CAT = catalase; GSH-Px = glutathione peroxidase. *Note:* GSH-Px is probably absent in *Drosophila*. (Adapted from Hauptmann, N. and Cadenas, E., in *Oxidative Stress and the Molecular Biology of Antioxidant Defenses*, Scandalios, J.G., Ed., Cold Spring Harbor Laboratory Press, Cold Spring Harbor, New York, 1997.)

oxidant in biological systems. It has a very short half-life, which means that its targets are in the direct environment of its formation, either in the bulk solution or in cellular structures, such as membranes and DNA (Yu, 1994). Since it is formed on a regular basis during normal cellular metabolic activities, the cell possesses both scavenging and repair systems to prevent or eliminate damage by OH•. The mechanisms of attack, damaging effects, and scavenging/repair system will be discussed here individually.

The hydroxyl radical oxidizes proteins, nucleic acids, lipids, and other cellular components. Oxidation of amino acids in proteins invariably leads to physical and chemical changes in these proteins. These changes include fragmentation, aggregation, and susceptibility to proteolytic digestion.

Oxidative damage leads to protein aggregation by denaturation of proteins. The aggregation could be related to the ability of hydroxyl radicals to form cross-linkages. Protein denaturation causes protein digestion by various proteolytic enzymes (Yu, 1994).

DNA damage induced by OH· includes both base alterations and strand breaks. Mitochondrial DNA mutates much faster than nuclear DNA, possibly because mitochondrial DNA is not covered by histones and is located near the mitochondrial respiratory chain, the richest cellular source of mutagenic ROS (Richter and Schweizer, 1997).

Oxidation of lipids is mainly related to peroxidation of polyunsaturated fatty acids of membrane phospholipids. Lipid peroxidation may be initiated by OH• or any other primary free radical of sufficient reactivity, by abstracting a hydrogen atom from the substrate (LH + free radical initiator → L•) (L stands for lipid). The presence of redox-active metals, such as copper and iron, can facilitate the initiation process. Propagation of lipid peroxidation relies on the interaction of molecular oxygen with carbon-centered free radicals to yield lipid peroxyl radicals (L• + O_2 → LOO•), which tends to capture labile hydrogen atoms of neighboring polyunsaturated lipids. This results in a chain reaction of lipid peroxidation, which propagates along membranes, yielding lipid hydroperoxides (L′H + LOO• → L′• + LOOH). It can be stopped by a termination reaction, such as the recombination of lipid peroxyl radicals (LOO• + L′OO• → LOOL′ + O_2), or by scavenging these radicals (Chaudère, 1994; Yu, 1994). The process of lipid peroxidation is summarized in Figure 4.2.

The autoxidation rate is highest in membranes with a high fraction of unsaturated lipids. Above a concentration of 10^{-6} M, hydroperoxides induce irreversible structural alterations that affect membrane fluidity and

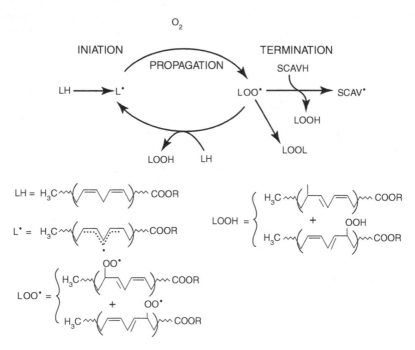

Figure 4.2 Autoxidation of polyunsaturated fatty acids in phospholipid membranes. During initiation, lipids (LH) are oxidized to lipid radicals (L•) by any free radical of sufficient reactivity. Propagation (a chain reaction) relies on reaction with oxygen (O_2), yielding lipid peroxyl radical (LOO•) and hydrogen abstraction from neighboring lipids, forming lipid hydroperoxides (LOOH). Termination occurs by recombination of lipid peroxyl radicals or by scavenging by scavengers in a reduced state (SCAVH), yielding lipid peroxide and a stable scavenger radical (SCAV•). (After Chaudère, J., in *Free Radical Damage and Its Control*, Rice-Evans, C.A. and Burdon, R.H., Eds., Elsevier Science, Dordrecht, the Netherlands, 1994. With permission.)

permeability and enzyme activities. Lipid hydroperoxides and peroxyl radicals oxidize sensitive amino acid residues, such as cysteine, thereby affecting enzymes in which such amino acids are essential. Mitochondria are strongly affected by lipid hydroperoxides, which are inhibitors of oxidative phosphorylation and which alter the cation permeability of their membrane (Chaudère, 1994). Decomposition of lipid hydroperoxides yields alkoxyl radicals (RO$^\bullet$). In the presence of iron (II), these alkoxyl radicals will often decompose through β-cleavage (cleavage of a carbon bond) into cytotoxic aldehydes (RCHO), for example, malonyldialdehyde (MDA) (Chaudère, 1994; Rice-Evans, 1994). These aldehydes are capable of forming protein cross-linkages that inactivate many cellular constituents, including membranes and enzymes (Yu, 1994).

The possible formation of all these radicals and reactive end products is basically a consequence of superoxide anion radical generation and processing and may occur, in principle, under normal cellular conditions. Remarkably, the same superoxide anion radical may also demonstrate opposite effects by acting as a terminator of lipid peroxidation reactions by recombination with lipid peroxyl radicals (Nelson et al., 1994). Therefore, the current view is that in a healthy cell an optimal balance exists between superoxide production and scavenging (McCord, 1995). Table 4.1 summarizes the derivatives discussed and their characteristics.

Table 4.1 Some characteristics of reactive oxygen derivatives

Species	Chemical symbol	Properties
Superoxide anion radical	$O_2^{\bullet-}$	Good reductant; poor oxidant
Singlet oxygen	$^1O_2$	Powerful oxidizing agent with half time of 10^{-6} s
Perhydroxyl radical	$HO_2^\bullet$	Stronger oxidant and more lipid soluble than superoxide
Hydrogen peroxide	H_2O_2	Oxidant with sluggish reactions; high diffusion capability
Hydroxyl radical	$OH^\bullet$	Extremely reactive; very low diffusion distance
Peroxyl radical	$ROO^\bullet$	Low oxidizing ability relative to $OH^\bullet$ but great diffusibility
Alkoxyl radical	$RO^\bullet$	Intermediate in its reactivity with lipid between $ROO^\bullet$ and $OH^\bullet$
Lipid hydroperoxide	$ROOH$	End product; decomposition yields $ROO^\bullet$ and $RO^\bullet$
Aldehyde	$RCHO$	Degradation product; cytotoxic; reacts with proteins
Nitric oxide radical	$NO^\bullet$	Oxidizes thiols; reacts with $O_2^{\bullet-}$ to form $ONOO^-$
Peroxynitrite	$ONOO^-$	Oxidizes thiols; decomposes into $OH^\bullet$

Source: Adapted from Yu, B.P., *Physiol. Rev.* 74: 139–162, 1994. With permission from the American Physiological Society.

To control the downstream effects of the production of these radicals within reasonable limits so that the cell remains in homeostasis, the cell possesses another set of scavengers and regeneration and repair enzymes.

First is the group of vitamins (A, C, and E). The principal biological lipid antioxidant is vitamin E, which can be taken to be represented by α-tocopherol (α-TH). Oxidation of α-TH to α-tocopheroxy radical (α-T$\cdot$) occurs when the antioxidant quenches a lipid peroxyl radical (LOO$\cdot$) or converts $O_2\cdot^-$ or OH$\cdot$ into less reactive species. The α-T$\cdot$ is regenerated by vitamin C (ascorbate) and probably by vitamin A (β-carotenoid). This fast regeneration probably forms the basis for a synergistic effect observed when all vitamins are available (Diplock, 1994). The regeneration process occurs at the lipid/water interface, since vitamin C is hydrophilic (Chaudère, 1994). Vitamin C is widely distributed in the cell and directly scavenges $O_2\cdot^-$ and OH$\cdot$ (Yu, 1994). Its oxidized form is enzymatically regenerated, at which point GSH is used as a substrate (Diplock, 1994). Vitamin C serves as both an antioxidant and a pro-oxidant. At high concentrations, it may act as pro-oxidant in the presence of oxidized transition metals, which in reduced form, promote lipid peroxidation (Yu, 1994).

To avoid decomposition of lipid hydroperoxides (LOOH) to reactive species, such as aldehydes (RCHO), or to prevent re-initiation of chain reactions by interference of reduced transition metals with lipid hydroperoxides, it is essential that lipid hydroperoxides be removed. The conventional GSH-P_x enzyme system described above is ineffective in reducing lipid hydroperoxides *in situ* in a membrane; the LOOH needs first to be released from the membrane by the action of phospholipase A_2 (PLA$_2$). However, the membrane-bound isoform of GSH-P_x (GSH-P_xm) differs in many respects from the conventional GSH-P_x. It is capable of catalyzing the reduction of LOOH *in situ* in its normal location in the membrane structure. This enzyme is thus more likely to confer protection against metal-catalyzed formation of radicals from LOOH than is the conventional GSH-P_x. The primary role of the conventional GSH-P_x is the reduction of H_2O_2 in the hydrophilic environment of the cytosol and mitochondrial matrix (Diplock, 1994).

Another very important scavenger is GSH. GSH, the tripeptide γ-glutamyl-cysteine-glycine, is the most abundant low-molecular-weight thiol present in the cell, generally with intracellular concentrations of 0.5 mM but sometimes as high as 10 mM. Reduced GSH is characterized by its reactive thiol group and, as an effective reductant, it plays an important role in a variety of detoxification processes. GSH readily interacts with free radicals and oxidizing compounds, such as H_2O_2, $O_2\cdot^-$, OH$\cdot$, and carbon radicals, including the nullification of lipid peroxide damage (Yu, 1994). GSH-mediated chain-breaking in the lipid peroxidation process results in the production of GSSG and $O_2\cdot^-$, at which point $O_2\cdot^-$ acts as a radical sink. The protection thus depends on the presence of SOD (Chaudère, 1994). GSH/GSSG ratios vary greatly in the different compartments of the cell. Mitochondrial GSH concentration is high; cells cannot survive acute depletion of GSH in this organelle. By contrast, cells can survive acute cytosolic GSH depletion (Chaudère, 1994). Apparently, 80 to 95% of ROS formed are normally scavenged by the mitochondrial GSH redox system. This scavenging by GSH may cause an imbalance in the thiol status and affect the Ca^{2+} homeostasis in the mitochondria (Goossens et al., 1995). These downstream effects of oxidative stress will be discussed in the next section.

A related phenomenon in maintaining cellular homeostasis is its redox state, which is mainly determined by the balance of reduced and oxidized thiols. ROS may affect thiols; some ROS that participate in alterations of this balance are the superoxide anion radical ($O_2^{\bullet-}$), peroxynitrite ($ONOO^-$), hydrogen peroxide (H_2O_2), and the hydroxyl radical ($OH^{\bullet}$). Inappropriate production or removal of these species leads to an alteration of redox homeostasis, which compromises the cell's ability to respond to additional insults (Flores and McCord, 1997; Wolin and Mohazzab-H, 1997). Downstream effects of a disturbance of cellular redox homeostasis will be discussed in the next section.

To maintain its preferred redox state, the cell activates its NADPH-dependent thioredoxin (TRX) enzyme system, which rapidly restores modified thiols to their unmodified reduced state (Wolin and Mohazzab-H, 1997). TRX not only functions as a proton donor for numerous proteins but probably also scavenges ROS (in particular H_2O_2 and $OH^{\bullet}$) and reactivates denatured proteins that contain mispaired disulfide bonds (Jacquier-Sarlin and Polla, 1996). Its role in reactivating transcription factors affected by oxidative stress will be discussed in the next section.

One final aspect affecting cellular homeostasis has to be reviewed. The previous discussion shows that the damaging effect of oxygen free radicals is highly dependent on the presence of a transition metal, of which iron is physiologically the most likely. The cell has a perfect mechanism to store this iron, that is, ferritin. Iron is stored in ferritin in the ferric (Fe^{3+}) form. At transport, Fe^{3+} is chelated to improve solubility. Iron can be released from ferritin, however, by reduction by the superoxide anion radical or by ascorbate. This release of the ferrous (Fe^{2+}) ion may lead to redox-cycling reactions resulting in the production of, for instance, $OH^{\bullet}$ radicals and in the initiation of lipid peroxidation (Chaudère, 1994; Deshpande and Joshi, 1985; Koster and Sluiter, 1994). Table 4.2 summarizes the most important antioxidant systems used by the cell to prevent oxidant injury and to maintain homeostasis.

4.2.2 Cellular responses to oxidative stress

The concentrations of individual ROS are usually tightly controlled by metabolizing and scavenging systems, as described above. Among other factors, an excessive production, owing to environmental agents and stressors, results in cytotoxic effects, which may lead to cell death. Oxidative stress is initiated by a variety of stimuli causing an array of (combined) effects, which may be related to inhibition of enzymatic activity, disturbance of metabolic processes, disturbance of signal transduction, and damage to DNA and cellular structures. Most of these effects are a consequence of an excessive production of the superoxide anion radical.

Table 4.2 Major antioxidant systems

Antioxidant systems	Primary localization	Actions
Cu/Zn SOD	Cytosol/nucleus	Catalyzes dismutation of $O_2^{\bullet-}$ to H_2O_2
Mn SOD	Mitochondrium	Catalyzes dismutation of $O_2^{\bullet-}$ to H_2O_2
CAT	Peroxisomes	Catalyzes reduction of H_2O_2 to H_2O
GSH peroxidases:		
GSH-Px	Cytosol/	Catalyzes reduction of H_2O_2 and other
GSH-Px,m	mitochondrium	hydroperoxides
	Lipid membranes	Catalyzes reduction of lipid hydroperoxides
GSH reductase	Cytosol/ mitochondrium	Catalyzes reduction of low molecular weight disulfides
TRX (Thioredoxin)	Cytosol/ mitochondrium/ nucleus	Restores redox state by reducing oxidized thiols, scavenges H_2O_2 and $OH^{\bullet}$, and repairs damaged proteins
α-Tocopherol (Vit. E)	Lipid membranes	Scavenges $O_2^{\bullet-}$, $OH^{\bullet}$ and $LOO^{\bullet}$; breaks lipid peroxidation chain reactions
β-Carotenoid (Vit. A)	Lipid membranes	Scavenges $O_2^{\bullet-}$; scavenges peroxyl radicals
Ascorbate (Vit. C)	Wide distribution (hydrophilic)	Scavenges $O_2^{\bullet-}$ and $OH^{\bullet}$; contributes to regeneration of vitamin E
GSH	Wide distribution	Substrate in enzymatic protection and GSH redox cycle; scavenges $O_2^{\bullet-}$, $O^{\bullet}$, and organic free radicals

Note: SOD = superoxide dismutase; CAT = catalase; GSH = glutathione; $O_2^{\bullet-}$ = superoxide anion radical; H_2O_2 = hydrogen peroxide; $OH^{\bullet}$ = hydroxyl radical; $LO^{\bullet}$ = lipid peroxyl radical. The GSH peroxidases have not yet been identified in *Drosophila*.

Source: Adapted from Yu, B.P., *Physiol. Rev.* 74: 139–162, 1994. With permission from the American Physiological Society.

Such production may primarily be brought about by compounds that inhibit the terminal enzymes of the mitochondrial respiratory chain, such as the insecticide rotenone, nitric oxide, cyanide, and the disinfectant azide (Richter and Schweizer, 1997). Heat shock also causes an increased production of $O_2^{\bullet-}$. According to Burdon et al. (1990), during heat stress at 42 to 45°C, an increased leakage of $O_2^{\bullet-}$ to the extracellular medium of mammalian cells was also observed. This leakage could be an outcome of changes in cellular membrane structures. Normally, $O_2^{\bullet-}$ molecules do not pass cellular membranes easily (Burdon, 1994). Surprisingly, however, excess $O_2^{\bullet-}$ inhibits CAT (Chaudère, 1994; Wolin and Mohazzab-H, 1997), notwithstanding that CAT would usually be a peroxisomal enzyme. Inhibition of CAT may lead to increased H_2O_2 concentrations in flies because they do not have a $GSH-P_x$ system to replace CAT.

Severe oxidative stress by $O_2^{\bullet-}$ generation may also be caused by redox cycling. Redox cyclers, such as the quinonoid compounds menadione and adriamycin, which form semiquinones, and other compounds, such as the herbicide paraquat, produce $O_2^{\bullet-}$ during the autoxidation of the reduced compound (Chaudère, 1994; Richter and Schweizer, 1997; Yu, 1994). (Referring to the description of the xanthine dehydrogenase pathway in the previous section,) two sources may strongly enhance the generation of $O_2^{\bullet-}$ via this pathway. First, a shift to anaerobic energy production activates this pathway by accumulation of AMP. Second, excessive Ca^{2+} release into the cytosol owing to oxidative stress leads to a conversion of the xanthine enzyme from the dehydrogenase form to the oxidase form. The consequence is a substantial increase of $O_2^{\bullet-}$ production (Winyard et al., 1994).

$O_2^{\bullet-}$ can be dismutated by SOD but also scavenged by GSH in mitochondria, yielding glutathione disulfide (Yu, 1994). Under severe oxidative stress conditions, this may lead to release of mitochondrial Ca^{2+}, disturbing Ca^{2+} homeostasis. Since mitochondria take up and release Ca^{2+} by separate routes, Ca^{2+} is cycled across the inner membrane, at which point ATP production is reduced by disturbance of the proton gradient over the inner membrane. Moreover, since Ca^{2+} release stimulates ROS production again and also regulates the activity of intramitochondrial dehydrogenases, as well as nucleic acid and protein synthesis, it is likely that disturbance of Ca^{2+} homeostasis in the mitochondrium may upset its functioning. This could be a starting point in the process leading to apoptosis (Fuchs et al., 1997; Goossens et al., 1995; Richter and Schweizer, 1997). The effects of excess $O_2^{\bullet-}$ production on mitochondrial homeostasis are summarized in Figure 4.3.

In addition to scavenging $O_2^{\bullet-}$, GSH also scavenges hydroxyl radicals (OH$^\bullet$) produced at the decomposition of H_2O_2 (Goossens et al., 1995; Jacquier-Sarlin and Polla, 1996). If this scavenging results in a change in cellular redox state owing to a drop in the GSH/GSSG ratio, signaling pathways are activated with downstream effects on transcription factors and gene expression. High concentrations of H_2O_2 resulting from excessive $O_2^{\bullet-}$ production may also lead to inhibition of glycolytic enzymes via changes in redox state and subsequent protein-S-thiolation (formation of mixed disulfides, RSSG, or RSSR', where R stands for a protein residue). This may partially inhibit glycolysis. The possible effects of H_2O_2 and their consequences for cellular homeostasis and viability are summarized in Figure 4.4 and discussed below in detail.

If the hydroxyl radical (OH$^\bullet$) is not efficiently scavenged by GSH (or TRX and vitamin C), lipid peroxidation may occur, leading to cytotoxic effects and finally to cell death. Lipid hydroperoxides may induce irreversible structural alterations that affect membrane fluidity and permeability and membrane-related enzyme activities caused by oxidation of sensitive amino acid residues, such as cysteine (Chaudère, 1994). In mitochondria, this may lead to inhibition of oxidative phosphorylation (Chaudère, 1994). Moreover, high cytosolic Ca^{2+} levels, owing to lipid peroxidation of membranes, can stimulate the lipolytic action of PLA_2 (Rice-Evans, 1994), thereby preferentially degrading the peroxidized form of phospholipids (Yu, 1994).

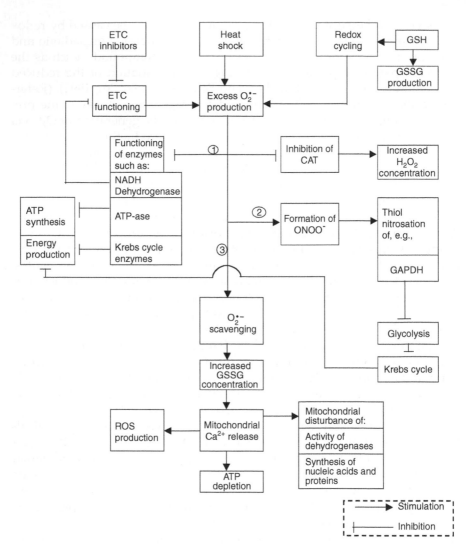

Figure 4.3 Effects of excess $O_2^{\bullet-}$ production in mitochondria. Apart dismutation of $O_2^{\bullet-}$ to H_2O_2 the three main effects are (1) enzyme inhibition, (2) thiol nitrosation, and (3) side effects of $O_2^{\bullet-}$ scavenging. See text for details. Abbreviations: $O_2^{\bullet-}$ = superoxide anion radical; ETC = electron transport chain; GSH = glutathione; GSSG = glutathione disulfide; NADH = nicotinamide adenine dinucleotide hydride; ATP = adenosine triphosphate; CAT = catalase; H_2O_2 = hydrogen peroxide; ONOO⁻ = peroxynitrite; GAPDH = glyceraldehyde-3-P-dehydrogenase.

Lipid peroxidation is a source of cytotoxic products, such as aldehydes, which are capable of forming protein cross-linkages that inactivate cellular constituents and enzymes (Yu, 1994). They may especially affect adenylate cyclase and phospholipase C activity, thereby influencing signal transduction

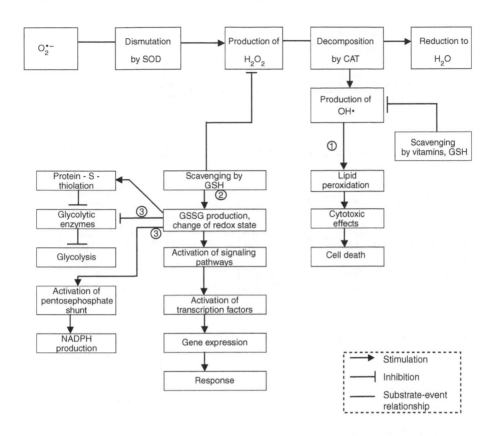

Figure 4.4 Three major side effects of H_2O_2 metabolism: (1) lipid peroxidation, (2) change of cellular redox state and activation of signaling pathways, and (3) deactivation of glycolysis and activation of the pentosephosphate (PP) shunt. Abbreviations: $O_2^{\bullet-}$ = superoxide anion radical; SOD = superoxide dismutase; H_2O_2 = hydrogen peroxide; CAT = catalase; OH$^{\bullet}$= hydroxyl radical; GSH = glutathione; GSSG = glutathione disulfide; NADPH = nicotinamide adenine dinucleotide phosphate hydride.

by the second messengers cAMP and Ca^{2+}, and cell proliferation (Chaudère, 1994; Rice-Evans, 1994). Perhaps the main peroxide-induced alterations are those that affect sodium and calcium homeostasis. Na^+/K^+-ATPase is strongly affected by lipid hydroperoxides. This implies that severe oxidative stress will usually be associated with cellular edema. Calcium homeostasis is also affected by an overload of lipid hydroperoxides. In some instances, this may be due to activation of voltage-dependent calcium channels in the plasma membrane, which lead to a sustained increase in intracellular calcium. Hydroperoxides also inhibit Ca^{2+}-ATPase pumps, which extrude calcium from the cytosol. This amplifies the build-up of cytosolic Ca^{2+} levels (Chaudère, 1994). All these effects of lipid peroxidation are summarized in Figure 4.5.

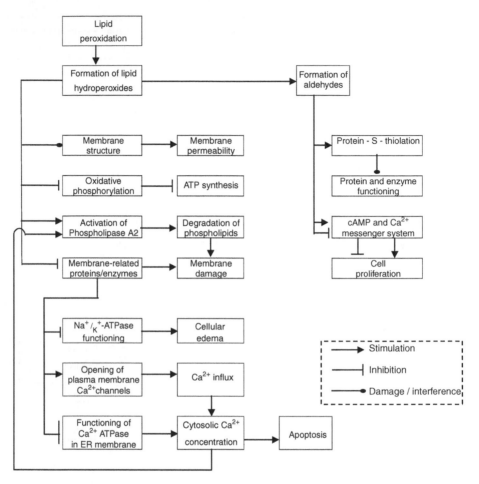

Figure 4.5 Important cellular effects of lipid peroxidation. These effects are related to structural integrity of the cell as well as cellular energy metabolism and Na^+/Ca^{2+} homeostasis, and may lead to cell death at severe stress. Cell proliferation may be stimulated or inhibited depending on the severity of the stress. Abbreviations: ATP = adenosine triphosphate; cAMP = cyclic adenosine monophosphate; ER = endoplasmic reticulum.

Figure 4.4 shows another major effect of H_2O_2 metabolism, that is, a change in redox state to an oxidizing environment owing to scavenging of H_2O_2 by GSH ($H_2O_2 + 2GSH \rightarrow GSSG + 2H_2O$). The average GSH/GSSG ratio may be close to 300 in some cell types, whereas, under severe oxidative stress conditions, this ratio can decrease to below 5. Thiol/disulfide exchange may couple the GSH redox buffer to the oxidation state of accessible cysteine residues in a number of proteins and enzymes. Such coupling reactions involve the reversible formation of intra- or intermolecular disulfides called protein-S-thiolation (P-SH + GSSG → PSSG + GSH) (Chaudère, 1994). At severe stress and high H_2O_2 concentration, levels of reducing agents (GSH,

TRX, vitamin C) may drop, and the NADPH buffer, required for the regeneration of the oxidized substances, may be depleted. As a consequence, the cell will stimulate NADPH production by activating the pentosephosphate shunt in the glycolysis. To provide sufficient glucose, glycogenolysis and gluconeogenesis are stimulated as well, whereas glycolysis is down regulated (Chaudère, 1994) (Figure 4.6).

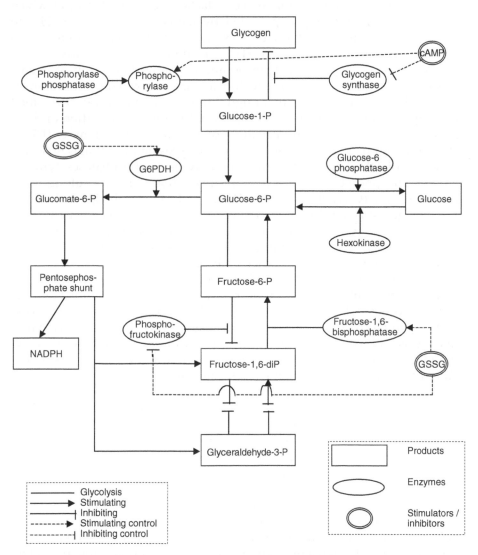

Figure 4.6 Part of the glycolysis pathway, showing stimulation of glycogenolysis, pentosephosphate shunt and part of gluconeogenesis, resulting in maximizing free glucose and NADPH levels. Shifts in processing are stimulated by increased GSSG and cAMP levels. Abbreviations: GSSG = glutathione disulfide; cAMP = cyclic adenosine monophosphate; NADPH = nicotinamide adenine dinucleotide phosphate hydride; G6PDH = glucose-6-phosphate dehydrogenase.

Figure 4.6 summarizes this change in glucose metabolism. The triggers for activating and deactivating the related enzymes are GSSG (leading to protein-S-thiolation) and cAMP, the level of which is increased by the shift in redox state.

If the mitochondrial respiratory chain is affected by stress, and NADH cannot unload its electrons and a shortness of NAD^+ arises, the cell shifts to an anaerobic glycolysis. In this process, lactic acid is produced, at which point the required energy is provided by the conversion of NADH into NAD^+, and the imbalance between these two components is restored. As a consequence, the Krebs cycle and ATP generation are compromised.

As shown in Figure 4.4, there is another important array of events caused by a shift in redox state. A shift toward an oxidizing environment may lead to activation of the basal signal transduction pathways described in Chapter 2. One common mechanism used to initiate activation of these signal pathways could be the activation of receptor tyrosine kinases (RTKs, shown in Figure 2.2) by inhibiting receptor tyrosine phosphatases (RTPs) with essential thiol groups. By inhibition of tyrosine phosphatases, tyrosine kinases phosphorylate themselves (autophosphorylation), resulting in an activated state (Guyton et al., 1996; Wolin and Mohazzab-H, 1997). Initiation of signal transduction by oxidative stress is more complicated, however: Different stressors may activate different pathways and different transcription factors. For example, H_2O_2 itself may activate the Ca^{2+}, cAMP, and MAPK systems, and in particular PKC (Burdon, 1994; Guyton et al., 1997; Wolin and Mohazzab-H, 1997). Moreover, PKC may influence the MAPK system either directly or indirectly via cAMP (see Figure 2.2) (Burgering and Bos, 1995; Guyton et al., 1997). Depending on the stress exerted by H_2O_2 and a combination of effects, it may lead to activation of enzymes in the classical pathway or the stress pathway of the MAPK cascade (Guyton et al., 1997; Holbrook et al., 1996). This may result in a different activation of various transcription factors, such as:

1. NF-κB, involved in cell activation and inflammatory response (Burdon, 1994; Winyard et al., 1994)
2. AP-1, which has a major role in proliferation (Burdon, 1994)
3. Heat-shock factor (HSF), leading to stress response by stress proteins (Jacquier-Sarlin and Polla, 1996)

Activation of the mammalian HSF-1 by oxidative stress is mediated by TRX, which maintains a certain redox state. At sufficient reducing power, HSF can be activated and then bind to the DNA promoter. At too-low TRX concentrations, HSF remains in an oxidized form and transcription is inhibited (Jacquier-Sarlin and Polla, 1996; Storz and Polla, 1996). TRX is induced, however, by H_2O_2; thus, it may result in a delayed response (Jacquier-Sarlin and Polla, 1996). NF-κB is localized in the cytosol in its inactive form, bound to inhibitor-κB (I-κB) (Storz and Polla, 1996). Probably, depending on the redox state, NF-κB is activated by dissociation from I-κB by a phosphorylation step. Activated NF-κB is then translocated to the nucleus for action (Baud and Karin, 2001). According to Schmidt et al. (1995), there is a correlation between accumulated H_2O_2 and activation of NF-κB. Thus, H_2O_2 would act as a messenger, probably conveying its message via the MAPK route or PKC. This also means that overcapacity of H_2O_2-decomposing enzyme systems could have a negative effect on other response systems (Schmidt et al., 1995; Storz and Polla, 1996).

Other stressors show different effects, however. For instance, heat shock, sodium arsenite, and osmotic stress exert their effects via the stress pathway of the MAPK cascade. These stressors activate MAPK-activated pro-tein-kinase 2 (MAPKAP-K2), an effector enzyme that phosphorylates and activates small heat-shock proteins (SHSPs) (Dérijard et al., 1995; Kyriakis and Avruch, 1996; Rouse et al., 1994), leading to stress response. However, phorbol esters, which activate PKC directly, exert their effect via the classical route of the MAPK cascade (Dérijard et al., 1995). The involvement of basal signal transduction systems in the oxidative stress response is summarized in Figure 4.7. Hydrogen peroxide, being both an environmental stressor and a major metabolite during normal cellular processing and under conditions of oxidative stress, is the central compound in this scheme.

It may be concluded that compounds such as GSH and TRX, which play a central role in maintaining the preferred cellular or compartmental redox state, are pivotal in cellular defense against oxidative stress. A shift in redox state toward an oxidizing environment has numerous consequences for enzyme activity, Ca^{2+} homeostasis, activation of basal signal transduction systems, and so on. The underlying mechanism for these processes is mainly oxidation of thiol groups.

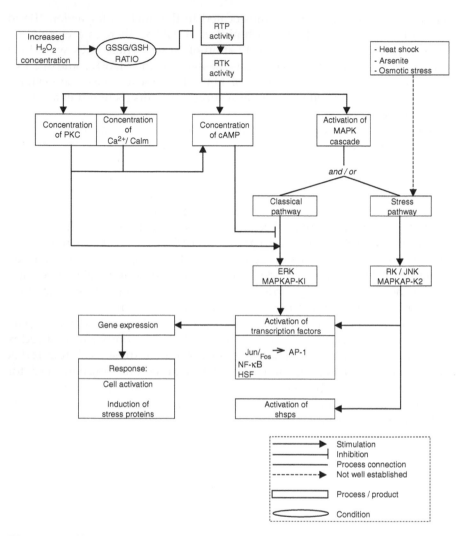

Figure 4.7 Effect of H_2O_2 on activation of signaling pathways. Depending on the severity of the stress and the redox state, H_2O_2 may activate different pathways, leading to a selective activation of transcription factors and response. Abbreviations: H_2O_2 = hydrogen peroxide; RTP = receptor tyrosine phosphatase; RTK = receptor tyrosine kinase; GSH = glutathione; GSSG = glutathione disulfide; cAMP = cyclic adenosine monophosphate, Calm = calmodulin; PKC = protein kinase C; MAPK = mitogen-activated protein kinase; ERK = extracellular signal-regulated kinase; JNK = c-Jun N-terminal kinase; RK = reactivating kinase; MAPKAP-K1/2 = MAPK-activated protein-kinase 1/2; AP-1 = activator protein-1; NF-κB = nuclear factor-κB; HSF = heat-shock factor.

chapter 5

The metallothionein system

5.1 Introduction

Metallothioneins (MTs) are small, cysteine-rich proteins that bind metal ions with high affinity. They have been conserved throughout evolution from yeast to man. It is generally accepted that the principal roles of MT lie in the regulation of the metabolism of essential trace metals, such as copper and zinc, and the detoxification of other heavy metals, such as cadmium, by means of sequestration. To maintain metal homeostasis, basal levels of MT are present in cells. These levels may vary depending on the type of cell or tissue, the life-stage of the animal, dietary fluctuations, and so on. Arthropods, especially terrestrial arthropods, are confronted with heavy metal pollution in soil, and they have therefore developed sequestration and excretion capacity. Insects, such as *Drosophila*, are also confronted with heavy metal pollution via dietary intake, and they have developed identical mechanisms to minimize the burden and develop tolerance. This subject will be discussed further in Section 7.3.2.

There is increasing evidence of another role for MT, that is, as a free radical scavenger (Bauman et al., 1991; Min et al., 1999; Sato and Bremner, 1993; Zhang et al., 2001). This role of assisting the oxidative stress defense system (see Chapter 9) emphasizes the function of MT as a member of an integrated cellular stress defense system.

5.2 Properties and types of metallothionein

In describing the functions of MT, Hamer (1986) looked at the individual metals and the affinity of MT for them. Hamer (1986) ascertained that (a) Cd^{2+} is highly toxic at high concentrations and nonessential, Cu^{2+}/Cu^+ is also toxic at high concentrations but has an enzymatic function, and Zn^{2+} is relatively nontoxic and essential for a large number of enzymes; (b) MT binds to each of them with a different affinity; the stability constant for copper is roughly a hundred times greater than that for cadmium, which in turn is roughly a hundred times greater than that for zinc. This leads to the postulation that MT has different functions for each of the metals. As the free copper ion is a potent toxin at high concentrations, Hamer (1986) proposed that the primary function of MT is to maintain a low level of copper ions while permitting enzyme activation. Considering the extremely long half-life of cadmium in organisms, the main function of MT in this respect is protection against long-term toxicity. Under normal conditions, cadmium is less abundant than copper, and basal levels of MT are sufficient to chelate low levels of Cd^{2+}. In the case of zinc, MT would act as a zinc-storage protein that plays a true homeostatic role in zinc metabolism. Jiang et al. (1998) have identified GSSG as a cellular ligand that reacts with MT and mobilizes zinc. This finding implies that the zinc content of MT is linked to the redox state of GSH in the cell in such a manner that zinc remains bound to MT as long as high thiol reducing power prevails. Thus, zinc is released once the redox balance becomes more oxidizing.

Theoretical support for the role of MT as a free radical scavenger is obtained when looking at the rate constants shown in Table 5.1. Indeed, _in vitro_ tests demonstrated that, on a molar basis, MT was 800 times more effective than GSH in preventing degradation of DNA by hydroxyl radicals. Under physiological conditions, however, MT concentration may be 1/200 of that of GSH, suggesting that the capacity of MT to scavenge hydroxyl radicals is comparable to that of GSH (Sato and Bremner, 1993).

Whereas MTs in mammals are classified in four groups ($MT^1/_{IV}$), each of which are present in various isoforms (Palmiter, 1998; Zhang et al., 2001), _Drosophila melanogaster_ synthesizes two MT forms: MTO, a 43-residue peptide that includes 12 Cys, and MTN, a 40-residue peptide that includes 10 Cys (Valls et al., 2000). These proteins display only 25% similarity (Bonneton et al., 1996). Both genes (Mto and Mtn) are expressed differently during development, and their expression can be increased by the inductive effect of various metals (Bonneton et al., 1996; Zhang et al., 2001). Domenech et al. (2003) report that both MTO and MTN can be characterized as copper MTs, because the coordination and folding of the MT is most stable in this configuration. They realize, however, that this classification does not imply the inability of these MTs to interact with other metal ions under intracellular conditions.

Table 5.1 Bimolecular rate constants for the reaction of hydroxyl and superoxide radicals with metallothionein (MT), glutathione (GSH), and superoxide dismutase (SOD)

	Hydroxyl radicals $K_{\cdot OH/MT, GSH, SOD}$ ($\times 10^9$ M^{-1} s^{-1})	Superoxide radicals $K_{O2^-/MT, GSH, SOD}$ ($\times 10^5$ M^{-1} s^{-1})
Zn (II)-MT	2.700	4 67
GSH	8	>10.000
SOD	–	

Source: After Sato, M. and Bremner, I., _Free Rad. Biol. Med._ 14: 325–337, 1993. With permission from Elsevier.

According to Bonneton et al. (1996), the function of MTO would be mainly related to active metal metabolism (maintaining essential levels) in the cell, while the MTN function would be more directly related to metal detoxification, probably in the lysosomes of the cell. This conclusion is in part based on a tissue-specific gene expression of these MTs. Both the Mto and the Mtn genes are mainly expressed in the digestive tract, but Mtn is also expressed in the fat body and Malpighian tubules. This activity could be due to the uptake by these organs of metal ions present in the surrounding hemolymph (Bonneton et al., 1996).

The intracellular allocation of tasks and place of MTO and MTN may be related to differential gene composition. The Mtn gene has two alleles, Mtn[1] and Mtn[3], and the Mtn[1] allele is sometimes duplicated. Strains containing duplicated Mtn[1] alleles synthesize Mtn mRNA at a higher level than those carrying a single Mtn[1] gene and tolerate increased metal concentrations (Bonneton et al., 1996). A remarkable difference in C-terminal composition possibly supports the different allocation of task and place: The Mtn[1] C-terminal has a glutamate (acidic) residue, whereas Mtn[3] and Mto have a lysine (basic) residue. This could lead to differentiation, with a function for the Mtn[1] product in the lysosome and for the Mtn[3] and Mto products in the cytoplasm. Maroni et al. (1995) relate the difference in C-terminal residues to differences in binding affinities and designate Mtn[1] as the one with the lower binding affinity, offsetting its higher level of expression.

Hensbergen (1999) made a comparison between MT from *Orchesella cincta* (Collembola) and MTO and MTN[1] from *D. melanogaster* (see Figure 5.1). Figure 5.1 provides an interesting view of this analysis. It shows why the identity between MTO and MTN is only 25%: MTO covers both N- and C-terminal parts of the *O. cincta* MT, whereas MTN covers only part of the N-terminal side of the *O. cincta* MT.

5.3 Regulation of metallothionein

Cells possess specific metal-responsive transcription factors (MRTFs) for activation of MT genes. In mammalian cells, six MRTF activities have been reported in the literature. The best characterized MRTF is the metal-responsive transcription factor-1 (MTF-1), an essential zinc-finger protein expressed

```
1        MVCKGCGTNCQCSAQK-----------CG----DNCACNKDC--QCVCKNGPKDQCCSNK
         :  :: ::.:..  .            ::    .. ..:  :  .:::::.:  .  ::.::
2    MSSTQGSASEAIRNCLCCGENCKCCGAEGKSPTCCKEKKCCGGGATQTASCCTCCGPDCVCKDGASLPCCANKTCCK
         ::.:  ..   ::  .:::::  .  ::   :  :
3   MPCPCGSGCKCASQATKGSCNCGCDCKCGG-DKKSACGCSE
```

Figure 5.1 Amino acid alignment of deduced amino acid sequences of metallothionein cDNAs from *Orchesella cincta* and *Drosophila melanogaster*. 1: *D. melanogaster* MTO, 2: *O. cincta*, 3: *D. melanogaster* MTN[1]. Identical residues (:), related residues (.). Identity between 1 and 2: 32.2%, 2 and 3: 46.7%, and 1 and 3: 25%. (After Hensbergen, P.J., Metallothionein in *Orchesella cincta*. Ph.D. thesis, Vrije Universiteit, Amsterdam, 1999.)

in all tissues (Latchman, 1995). In resting cells, it is localized in the cytoplasm, while it is transferred to the nucleus under stressful conditions (Zhang et al., 2001). Recently, Zhang et al. (2001) characterized the *Drosophila* homolog for vertebrate MTF-1, dMTF-1, which is most similar to its mammalian counterpart in the DNA-binding zinc-finger region. Yet some aspects of heavy metal regulation have been subject to divergent evolution between *Drosophila* and mammals. These differences will be discussed further below.

How the MT genes are transcriptionally activated is still puzzling (Domenech et al., 2003). Whether the MRTFs perform distinct metalloregulatory functions is not known. The factors seem to influence each other in competing for MT promoter binding, which may even lead to a positive or negative control of transcriptional activation in a concentration-dependent manner (Koropatnick and Leibbrandt, 1995). It is a fact that MT genes from yeast to man contain metal-responsive elements (MREs) in their promoters. For instance, the *Drosophila* Mto gene contains four MREs and the Mtn gene, two MREs (Bonneton et al., 1996). According to Koropatnick and Leibbrandt (1995), there are three possibilities for activation of MRTFs and interaction with multiple MREs: (a) multiple MRTFs exist to respond to different metals; (b) a single MRTF exists with the ability to respond to several different metals (a "flexible" protein model described by Thiele, 1992); and (c) DNA-binding MRTFs exist that are specifically responsive to zinc only (a model presented by Palmiter, 1994). MTs, associated with zinc, release zinc by displacement upon exposure to either copper or cadmium. Released zinc would then be available to mediate MRTF activity, especially if the newly released zinc were present in the nucleus. MT has, indeed, been localized to the nucleus during development (Koropatnick and Leibbrandt, 1995). The theoretical background for this "free zinc" model is provided by Hamer (1986), based on the rank order of affinities between MT and copper, cadmium, and zinc.

The regulation of induction or degradation of MT on the basis of free zinc levels has been proposed by various investigators (Cherian, 1995; Giedroc et al., 2001; Karin, 1985; Koropatnick and Leibbrandt, 1995; Palmiter, 1994; Roesijadi, 1996). Roesijadi (1996) developed a hypothetical model (shown in Figure 5.2) for MT regulation on the basis of free zinc concentrations.

The model shown in Figure 5.2 covers a broad field of action, not only by heavy metals but also by other environmental stressors that directly or indirectly release zinc from ligands such as enzymes. This means that it is not only applicable to metal metabolism under homeostatic conditions, but that it could also be used for determining regulation of response to metal stress (discussed below) and oxidative stress. The model is not undisputed, however. Beyersmann and Hechtenberg (1997) reported, for instance, that the MRE-binding factors MRE binding factor-1 (MBF-1) and metal-element binding protein-1 (Mep-1), both from mouse cells, bind to MT promoters in a zinc-dependent manner but do not respond to cadmium. The metal-responsive transcription factor (MTF-1) in HeLa cells is induced, however, by zinc or cadmium but is only moderately activated *in vitro* by toxic concentrations of these metals (Beyersmann and Hechtenberg, 1997). Dalton et al. (1997)

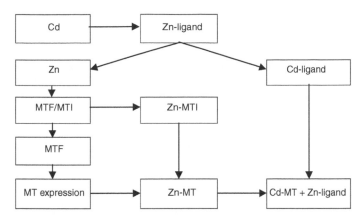

Figure 5.2 Model for coupled MT induction and rescue of target ligands, compromised by inappropriate metal binding, by cadmium in this example. Abbreviations: MT = metallothionein; MTF = metal-responsive transcription factor; MTI = metallothionein transcription inhibitor (After Roesijadi, G., *Comp. Biochem. Physiol.* 113C: 117–123, 1996. With permission from Elsevier.)

observed an *in vitro* DNA-binding activity of MTF-1 in mouse cells by zinc or oxidative stress-inducing agents, but according to Andrews et al. (1999), cadmium had no effect on the amount of MTF-1 binding in Hepa cell lysates. In *Drosophila* cells, dMTF-1 is more activated by copper or cadmium than by zinc (Domenech et al., 2003). In fact, dMTF-1 is unable to induce transcription in response to zinc at concentrations that readily induce transcription in mammalian cells (Zhang et al., 2001). One conclusion of Beyersmann and Hechtenberg (1997) therefore is that different MRE-binding proteins respond to different metal ions in a concentration-dependent manner. This conclusion would support alternative (a) of the above-mentioned proposal by Koropatnick and Leibbrandt (1995) rather than the model of Roesijadi (1996) shown in Figure 5.2 and described by alternative (c) above. It is important to remember that the model described by Roesijadi (1996) may not be uniformly applicable or that *in vivo* regulation is more complicated than the model shows.

The differences between MTF-1 and dMTF-1 are mainly related to metal and tissue specificity. MTF-1 is predominantly activated by zinc or cadmium, whereas dMTF-1 is activated by copper or cadmium. According to Zhang et al. (2001), this cannot be attributed to a species difference between MTF-1 factors themselves, since dMTF-1, upon transfection into mouse cells, mediates zinc-responsive transcription like human MTF-1. These differences relate rather to differences in metal metabolism, such as tissue-specific sequestration (see Chapter 7) or export of heavy metals. For instance, the recently characterized cadmium-responsive protein CDR-1 (Liao et al., 2002) in the intestinal tract of *Caenorhabditis elegans* seems to be an integral lysosomal membrane protein that probably transports cadmium into the lysosome either for osmoregulatory purposes or for detoxification. In this respect, it is worth noting that the *Drosophila* genome also contains six homologs to mammalian zinc transporters (Zhang et al., 2001). This could help explain the second difference between the functioning of the transcription factors, that is, tissue specificity. Mammalian MTF-1 is expressed in all tissues analyzed so far, whereas dMTF-1 is expressed in a few tissues, notably gut and fat body (Zhang et al., 2001). The relationship of this aspect with tissue-specific sequestration is likely.

If *Drosophila* are subjected to heat shock, Mto and Mtn genes are expressed, but only in regions accumulating metals, such as the iron and copper cell regions in the midgut and the posterior midgut (Bonneton et al., 1996). Tests in the nematode *C. elegans* showed comparable results, leading to the assumption that one of the effects of heat shock might be disruption of metal binding to endogenous metallomolecules. Metal ions would thus be free to interact with the metal regulatory system of MT genes (Bonneton et al., 1996; Freedman et al., 1993). It is not unlikely that zinc, with its low stability constant (compared to copper and cadmium), would be the first metal released from enzymes and MTs by heat shock. In that case, zinc could play its role in inducing MTs as described in the Roesijadi model.

Cell- and tissue-specific gene expression of MTs points to differential regulation of expression that is metal related and concentration dependent. Metal-related expression is best exemplified in snails. Dallinger et al. (1997) claim the existence of specific MTs for cadmium and copper, in *Helix pomatia*. Moreover, these isoforms of MT seem to be tissue specific. Dallinger et al. (1997) have isolated the cadmium isoform from the midgut gland and the copper isoform from the mantle. It is unlikely that the Roesijadi model would be applicable in this form to activate these specific genes. A more complicated system is unavoidable for metal-specific MT induction. Cadmium, for instance, has not only the capacity of replacing zinc and activating transcription factors via the "zinc route," but is also able to activate protein kinase C (PKC), which may then phosphorylate transcription factors directly (Chen et al., 1999). Moreover, cadmium is also able to activate the Ca^{2+}/calmodulin and cAMP/PKA second messenger systems, as well as the MAPK pathway via plasma membrane receptors. Details of the mechanisms will be discussed below, but what matters here is that the transcription factor AP-1 and other MRTFs can be differentially activated without involvement of zinc release. AP-1 binds to the promoters of the Mtn and Mto genes, expressing MT in *D. melanogaster* (Bonneton et al., 1996). The antioxidant responsive element (ARE), to which AP-1 may bind, has been identified in the promoter of the mouse MT-1 gene (Andrews et al., 1999; Dalton et al., 1996) and in the Cdr-1 gene of *C. elegans* (Liao et al., 2002). In view of the fact that AP-1 also binds to Mtn/Mto genes of *Drosophila* it is not unlikely that the ARE would also be present on the promoters of these genes. In that case, oxidative stress effects could be transformed in MT gene expression via activation of MAPK and AP-1 and binding of the latter to AREs on MT promoters. DeMoor and Koropatnick (2000) as well as Wang and Templeton (1998) are of the opinion that MAPK and additional transduction pathways are involved in the response to cadmium ions.

On the basis of this analysis, a speculative model (shown in Figure 5.3) for the induction of *Drosophila* MTs and probably MTs of other invertebrate species, activated by copper, zinc, or cadmium, has been developed.

The model in Figure 5.3 provides a solution for the problem of metal-specific gene expression. It associates the activation of transcription factors by free zinc (route 1) with copper/zinc homeostasis, although this route is also available for cadmium. It is more likely, however, that cadmium uses the other routes (2 to 4) as well by exerting its specific reactions. For metal-specific gene expression, it is believed that there are multiple MRTFs that respond to different metals. In this way, the model is a combination of the alternatives (a) and (c) proposed by Koropatnick and Leibbrandt (1995). The model does not provide a solution for cell- and tissue-specific MT induction. This aspect seems to be more related to metal uptake, metabolism, and export. The model does not show the "rescue of target ligands" as in the Roesijadi model. The principle of multiple pathway regulation shown in Figure 5.3 is supported by the observation of Zhang et al. (2001) that the presence of AP-1 binding sites on the promoters of MT genes of *Drosophila* points to a possible interconnection between dMTF-1 and AP-1 in the cellular stress response of insects.

5.4   Heavy metal stress

Based on the above discussion, it may be clear that homeostasis of essential metals is a dynamic equilibrium that can be disturbed by potent toxic metals, such as cadmium. However, transition metals, such as copper, follow a different pathway in developing metal stress than divalent metals, such as cadmium. Both metals will therefore be used for describing what happens if cells are exposed to levels causing metal stress.

Cadmium ions are taken up through voltage-gated calcium channels in the plasma membrane of various cell types and accumulate intracellularly by binding to cytosolic and nuclear material (Beyersmann and Hechtenberg, 1997). Depending on its concentration, cadmium may cause a variety of effects and responses, ranging from induction of MT, GSH, proto-oncogenes, and stress proteins to inhibition of zinc-containing enzymes, generation of oxidative stress and lipid peroxidation, inhibition of ATPases leading to disturbance of Ca^{2+} homeostasis, disturbance of protein synthesis, and inhibition of DNA repair enzymes leading to genotoxic stress. The consequential damage to cellular functioning may lead to cell death. The effects, responses, and underlying mechanisms will now be discussed in more detail.

The basis of the toxicity of many metals, including cadmium, in biological systems is the formation of complexes with nucleophilic ligands of target molecules. The affinity of cadmium for numerous ligands under physiological conditions has the following order: thiol > phosphate > chloride > carboxyl. The affinity of cadmium is higher for biomolecules containing more

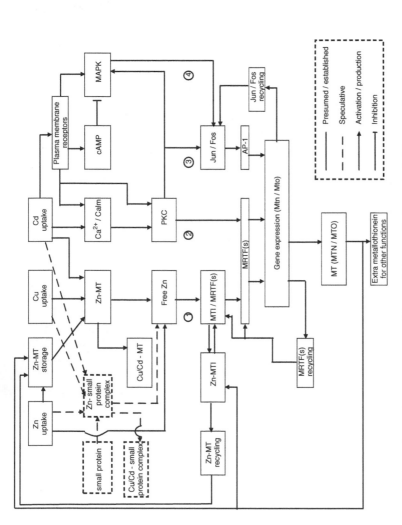

Figure 5.3 Speculative model for independent activation of *Drosophila* metallothionein (MT) genes by zinc, copper, or cadmium. Route 1 is the pathway leading to cellular copper/zinc homeostasis. Routes 1 through 4 may be used by cadmium to activate transcription factors. Route 2 leads to activation of MRTFs via PKC, route 3 to activation of Jun/Fos via PKC, and route 4 to activation of Jun/Fos via MAPK. The model provides a system for metal-specific gene expression. Discrimination between the Mtn and Mto genes depends on the specificity of the MRTFs involved. These are not yet known in *Drosophila*. Abbreviations: MT = metallothionein; Calm = calmodulin; cAMP = cyclic adenosine monophosphate; PKC = protein kinase C; MAPK = mitogen-activated protein kinase; MTI = metallothionein transcription inhibitor, MRTF = metal-responsive transcription factor; AP-1 = activator protein-1; Jun/Fos = components of AP-1; Mtn/Mto = genes in *Drosophila* expressing the MTs: MTN and MTO.

than one binding site, for example, MT (Goering et al., 1995). Another underlying factor of its toxicity is the fact that cadmium, on the basis of the rank order for substitution drawn up by Hamer (1986), replaces zinc in enzymes, thereby inhibiting their activity.

These factors together explain why even at low, noncytotoxic cadmium concentrations (1 to 5 μM), MT and GSH are induced (Beyersmann and Hechtenberg, 1997), probably via the pathways shown in Figure 5.3 and via redox regulation, respectively. A side effect of increased MT and GSH levels is that they provide tolerance for subsequent cadmium exposure (Beyersmann and Hechtenberg, 1997; Goering et al., 1995).

Perhaps the most unexpected and least understood genetic effect of cadmium is its ability to stimulate the expression of the proto-oncogenes c-jun, c-fos, and c-myc, also termed immediate early or primary response genes, and some other genes related to mitogenic stimulation (Beyersmann and Hechtenberg, 1997; Wang and Templeton, 1998). This is of interest because the gene products Jun and Fos form the heterodimeric transcription factor AP-1, shown in Figure 5.3. The way in which these proto-oncogenes are induced is disputed, however. Beyersmann and Hechtenberg (1997) reported that cadmium induced the proto-oncogenes by activation of PKC and probably the mobilization of intracellular calcium ions. Activation of the transcription factors by PKC seems to be direct, without involvement of the MAPK cascade and the cAMP route. Wang and Templeton (1998) demonstrated that Cd^{2+} induced c-Fos mainly by activation of the MAPK cascade. Yu et al. (1997) demonstrated that cadmium-induced MT gene expression in Chinese hamster cells also requires activation of PKC. They also showed that involvement of the AP-1 transcription factor was not necessary and that PKC probably phosphorylates the MTF-1 transcription factor directly. Their findings might support the Roesijadi model (Figure 5.2), but an extra phosphorylation step via PKC would be required for activation of MT gene expression. Anyhow, it is of interest to ascertain if a mitogenic pathway is involved in transduction of signals from metal stress. The activation of this pathway may be initiated by binding of Cd^{2+} to receptors on the plasma membrane or to calmodulin, simulating calcium binding in this way (Beyersmann and Hechtenberg, 1997; Goering et al., 1995). Although expected, Cd^{2+} binding to calmodulin could not be demonstrated by Wang and Templeton (1998).

By binding plasma membrane receptors, cadmium stimulates release of calcium from intracellular storage sites (Goering et al., 1995). Moreover, elevated cadmium levels may inhibit Ca^{2+}-ATPase working in the plasma and ER membranes, leading to disturbance of calcium homeostasis (Beyersmann and Hechtenberg, 1997; Goering et al., 1995). This may be an important effect caused by cadmium exposure, which may strongly influence overall cellular functioning and may initiate the process leading to apoptosis (described in Chapter 8).

In contrast to the induction of MT and GSH by low micromolar concentrations (1 to 5 μM), relatively high doses of cadmium, close to cytotoxic levels (10 to 50 μM), are required to induce most stress and other related proteins. The expression of these proteins thus may serve an emergency function, because they might rescue sensitive proteins from damage (Beyersmann and Hechtenberg, 1997). Cadmium-induced stress protein synthesis prior to cytotoxicity has also been observed in *Drosophila* (Goering and Fisher, 1995). A mechanism of metal-induced stress protein synthesis may involve proteotoxicity, since aberrant or denatured proteins are a stimulus for stress protein synthesis. A second mechanism of metal-induced stress protein synthesis may involve depletion of intracellular thiol pools, since several sulfhydryl reactive agents increase synthesis of these proteins (see Chapter 4). This mechanism includes thiol-containing enzymes as well (Goering et al., 1993; Goering and Fisher, 1995). Bauman et al. (1991) emphasize that the promoter of the human hsp70 gene contains a sequence that is homologous to the core of the human MT II MRE. Thus, upon cadmium exposure in human cells, hsp70 could be induced directly. Whether this is also the case in lower animals, such as *Drosophila*, is not known. If this is the case, the speculative model of Figure 3.20 would require adjustment. As described above, however, activation of stress protein gene transcription requires higher cadmium concentrations than MT gene expression does. However, cadmium may also disrupt cellular calcium homeostasis, which may lead to induction of the so-called glucose-regulated proteins (GRPs) belonging to the stress proteins (see Chapter 3) (Goering et al., 1993; Goering and Fisher, 1995).

At cytotoxic levels, cadmium inhibits enzymes, such as superoxide dismutase (SOD) and catalase (CAT), involved in the detoxification of reactive oxygen species (ROS) by displacing essential metal cofactors, such as zinc, or by binding to essential thiol groups. This may lead to increased levels of O_2^- and generation of OH', resulting in lipid peroxidation and oxidative stress (Beyersmann and Hechtenberg, 1997; Goering et al., 1995; Kawanishi, 1995). The induction of MT described above is independent of these oxidative stress effects (Cherian, 1995). The same applies to the induction of stress proteins by cadmium; they are induced prior to cytotoxic effects caused by oxidative stress. On the contrary, at cytotoxic cadmium levels, *de novo* synthesis of proteins in general is impaired (Beyersmann and Hechtenberg, 1997). After chronic cadmium dosing for 6 months, decreases in hepatic mixed-function oxidase activities and cytochrome P450 content have also been demonstrated. The reduction in cytochrome P450 is due to increased heme degradation, which results from cadmium stimulation of heme oxygenase activity (Goering et al., 1995).

Parallel to cytotoxic effects, such as lipid peroxidation, inhibition by cadmium of DNA repair enzymes, such as DNA polymerases, may take place (Beyersmann and Hechtenberg, 1997). As possible mechanisms of repair inhibition, either the direct interaction of cadmium with thiol groups of repair enzymes or the interaction with calcium-regulated processes has been suggested (Kawanishi, 1995). The cadmium ion may also bind to isolated nucleic acids and chromatin, causing conformational changes at very high concentrations of 1 mM or higher (Beyersmann and Hechtenberg, 1997).

DNA strand breaks observed during cadmium exposure at high concentrations seem to be an indirect effect. It has been noticed that ROS generated by cadmium exposure, for example, by inhibiting detoxifying enzymes, such as SOD, cause the strand breaks (Beyersmann and Hechtenberg, 1997; Kawanishi, 1995). However, the ROS generated by cadmium exposure may induce MT, which depending on its subcellular localization, may provide some protection against DNA damage (Schwartz et al., 1994). Min et al. (1999), however, found that MT might have opposite effects on DNA in nuclei, independent of metals bound to MT. At high concentration, MT can induce DNA strand scission by itself (Min et al., 1999). A possible mechanism could be the release of reduced transition metal from MT and local generation of hydroxyl radicals in a Fenton-driven reaction. This could be, for example, reduced copper ion (Cu^+), which binds to MT. This will be further analyzed below.

When copper enters a cell, it seems to be bound to GSH first as a Cu^+–GSH adduct (Brouwer and Brouwer-Hoexum, 1998; Jiménez et al., 2002). According to tests with the blue crab, *Callinectes sapidus*, by Brouwer and Brouwer-Hoexum (1998), copper then binds to newly synthesized MT with a concomitant decrease in Cu^+–GSH. It is generally known, however, that copper can substitute zinc in Zn–MT, whereby zinc could induce the synthesis of new MT (Hamer, 1986; Fabisiak et al., 1999). In a fully reduced environment, MT is saturated with copper in a Cu_{12}–MT configuration (Fabisiak et al., 1999; Oikawa et al., 1995). In a more oxidizing environment, for example, in the presence of H_2O_2, some copper can be released, which may then participate in redox reactions. MT fully saturated with copper is therefore considered a pro-oxidant (Jiménez et al., 2002; Stephenson et al., 1994). MT with less than 6 mol Cu^+/mol MT is a stable configuration (Fabisiak et al., 1999). Liberated Cu^+ may participate in the Fenton reaction to yield OH^-. Another thermodynamically feasible pattern is as follows: Cu^+ reacts with H_2O_2 to give Cu^{2+}. This ion reacts with GSH to give Cu^+ and a thiyl radical, $GS^•$. This radical reacts with GS^- to give $GSSG^{•-}$. The latter is a strongly reducing species that reacts rapidly with oxygen to yield $O_2^{•-}$ (Brouwer and Brouwer-Hoexum, 1998). Part of this reaction path is also followed if copper enters a cell as Cu^{2+}.

Copper–MT in copper-loaded cells does not cause lipid peroxidation, as long as the copper–MT is in a sufficiently reducing environment with high levels of GSH. In the presence of oxidizing agents, however, such as H_2O_2, copper–MT can enhance the process of lipid peroxidation initiated by oxidizing agents (Jiménez et al., 2002). In this respect, copper as a transition metal exhibits different toxic properties than cadmium and zinc in cadmium/zinc–MT (Roméo et al., 2000). Roméo et al. (2000) not only refer to the pathway leading to lipid peroxidation, but also emphasize that cadmium inhibits, for instance, the important enzyme catalase, whereas copper does not. As mentioned above, cadmium also may cause lipid peroxidation, but the sequence of events is different. Zinc does not evoke lipid peroxidation (Arthur et al., 1987).

Copper–MT may also damage DNA (Fabisiak et al., 1999). The type of damage and the means of chemical attack depends on the degree of saturation of MT by copper (Oikawa et al., 1995). Fully saturated copper–MT (Cu_{12}–MT) releases Cu^+ in an oxidizing environment, which, in the presence of H_2O_2, is able to oxidize thymine residues and to cleave the DNA backbone in an oxidative way by abstracting a hydrogen atom from a deoxyribose ring. MT loaded to a lesser extent with copper, for example, Cu_2Cd_5–MT, only causes backbone cleavage by an interaction between MT and DNA. Cleavage of the phosphate–sugar backbone then occurs in a nonoxidative way. Also, in this reaction mechanism, copper remains necessary, because Cd–MT is not able to cause the same effect (Oikawa et al., 1995).

In conclusion, metal stress by cadmium highly depends on intracellular levels of cadmium. At noncytotoxic cadmium levels, the cell responds by inducing MT to sequester free cadmium, as well as by inducing GSH to maintain redox state and stress proteins to prevent or repair damage. At high, cytotoxic levels, cadmium may disturb cellular homeostasis by interfering with signal transduction systems and inhibiting enzyme activity. At very high concentrations, cadmium causes oxidative stress, resulting in cytotoxic and genotoxic effects, which may lead to cell death. In the case of a transition metal, such as copper, the final result may be the same, but the pathway along which this result is achieved may differ.

chapter 6

The mixed-function oxygenase system

6.1 Introduction

Endogenous compounds, such as steroid hormones, are metabolized in the cell, either for activation or deactivation purposes (Waxman, 1988). These biotransformation processes are catalyzed by so-called cytochrome P450 enzymes, a diverse class of enzymes belonging to the mixed-function oxygenase (MFO) system. Cytochrome P450 enzymes are membrane-bound proteins, primarily localized in the microsomal fraction of the endoplasmic reticulum (microsomes). Cytochrome P450 enzymes have also been demonstrated in mitochondria (Feyereisen, 1999).

Environmental pollutants, such as insecticides and other xenobiotics, are also metabolized by the same enzymatic system. Metabolism of these compounds is directed to deactivation and excretion. In view of the great variety of xenobiotics degraded by the MFO system, this system is also highly specialized in insects. According to Feyereisen (1999), the insect genome probably carries about 100 P450 genes expressing different P450 (iso) enzymes to fulfill their detoxifying role. Induction or constitutive overexpression of specific enzymes may lead to tolerance to pollutants, such as specific insecticides. This will be discussed in Section 7.3.1. This chapter will concentrate on the description and regulation of the MFO system and the degradation mechanisms for xenobiotics in insects.

6.2 Description of the mixed-function oxygenase system

Biotransformation processes directed to deactivation and excretion of both endogenous and exogenous compounds take place in two phases. During phase 1, hydrophobic compounds are converted into hydrophilic intermediates by oxidation, reduction, and hydrolysis (Timbrell, 1991). During phase 2, these intermediates are made more water soluble by conjugation to facilitate excretion. In higher animals, these hydrophilic compounds are conjugated with compounds such as glucuronides, sulphates, and glutathione.

Although biotransformation of foreign compounds is directed to detox-ification, this is not always the case. Metabolites may be produced that are more toxic than their parent compounds. This may take place during both phase 1 and 2 reactions. This process is known as bioactivation (Timbrell, 1991). This aspect will get special attention in the following section, since it influences other cellular defense systems and demonstrates the interrelation-ships between these systems and the MFO system. Timbrell (1991) gives an overview of possible metabolic pathways available for a foreign compound. Some of these pathways may be detoxification pathways, while others may lead to toxicity. (summarized in Figure 6.1).

At the induction of enzymes of the MFO system, two pathways are distinguished. This differentiation is important because foreign compounds may induce different iso-enzymes of the MFO system and possibly cause different effects. If possible, foreign compounds are therefore divided into two classes: phenobarbital-type (PB-type) inducers and 3-methylcholan-threne-type (3MC-type) inducers. These classes are named as such, because PB and 3MC are often used as model substrates (van Straalen and Verkleij, 1991)

Induction of cytochrome P450 iso-enzymes by compounds of the 3MC-type takes place by binding of these compounds to an aryl hydrocarbon receptor in the cytoplasm. This substrate binding results in an activated

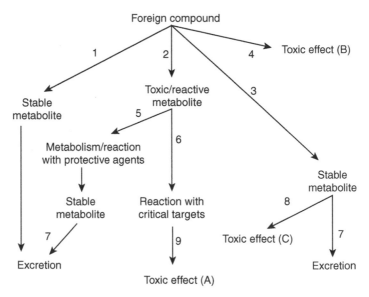

Figure 6.1 The various possible consequences of metabolism of a foreign compound. The compound may undergo (1) detoxification, (2) bioactivation, (3) formation of a stable metabolite that may cause a toxic effect (C), or (4) cause a direct toxic effect (B). The reactive metabolite may be detoxified (5/7) or cause a toxic effect (A) (6/9). (After Timbrell, J.A., *Principles of Biochemical Toxicology*, 2nd ed., Taylor & Francis, London, 1991. With permission.)

receptor complex, which gives rise to expression of special genes and synthesis of special iso-enzymes of the MFO system, such as cytochrome P450 IA1. This important iso-enzyme is also known as aryl hydrocarbon hydroxylase (AHH) (Timbrell, 1991). It is also of interest that the Ah receptor is stabilized by the stress protein hsp90 (hsp83 in the case of *Drosophila melanogaster*) in a manner identical to the steroid hormone receptor shown in Figure 3.4. (Gonzalez et al., 1993; van der Oost, 1998). If a 3MC-type compound binds to the Ah receptor, the complex is translocated to the nucleus by association with the Ah receptor nuclear translocator (Arnt). During this cascade of events, the complex is phosphorylated by protein kinase C (PKC). In the nucleus, the complex (or part of it) binds to the Ah-regulatory element (AhRE) on the promoter of specific P450 genes. The AhRE is related to the xenobiotic responsive element (XRE) (Gonzalez et al., 1993; Nebert et al., 2000). According to Jaiswal (1994), this XRE is found on some promoters of both phase 1 and 2 enzymes of the MFO system. On the contrary, the antioxidant responsive element (ARE), mentioned in Chapter 4, has only been found on promoters of phase 2 P450 genes. More recent findings are discussed below.

Compounds of the PB-type predominantly induce the cytochrome P450 IIB1 iso-enzyme (Goeptar, 1993; Timbrell, 1991), but PB also induces enzymes of the cytochrome P450 III family, which are normally induced by steroid hormones (Timbrell, 1991). The method of induction of cytochrome P450 enzymes by compounds of the PB-type inducers is different from the 3MC-type inducers: No receptor has yet been isolated for PB-type inducers (Gonzalez et al., 1993). One of the possible control mechanisms proposed by Timbrell (1991) is based on substrate–enzyme binding and subsequent feedback control by means of a repressor mechanism. Such a control system might also account for the enormous variety of chemical structures that act as PB-type inducing agents and for which it would be difficult to envisage a single receptor (Timbrell, 1991; Whitlock and Denison, 1995). More recent studies reveal that PB also stimulates the DNA binding of the transcription factor AP-1 (Jun/Fos) and activates gene expression of, for example, quinone reductase enzymes in an AP-1-dependent manner. Because the DNA-binding activity of AP-1 (Jun/Fos) appears to be regulated by the redox state of the cysteine residues in the DNA-binding domains of the two proteins (Jun/Fos), these findings suggest that PB might activate AP-1 by increasing intracellular oxidant levels, thereby activating gene expression indirectly (Whitlock and Denison, 1995). Indeed, Pinkus et al. (1995) report that the quinone adriamycin, which behaves as a PB-type inducer of P450 enzymes, generates hydroxyl radicals, causing oxidative stress and activation of the AP-1 transcription factor.

Compounds of the 3MC-type include polycyclic aromatic hydrocarbons (PAHs) and halogenated polycyclic hydrocarbons, such as polychlorodibenzo-p-dioxins (PCDDs) and co-planar polychlorinated biphenyls (PCBs) (Landers and Bunce, 1991; Krishnan and Safe, 1993). These groups of lipophilic chemicals have a rather planar structure in common, by which their binding specificity differs from, for example, steroid hormones. For instance, tetrachlorodibenzo-p-dioxin (TCDD) binds with high affinity to the Ah receptor but has no affinity for steroid hormone receptors. On the contrary, steroid hormones with a rather bent structure do not have much affinity for the Ah receptor (Landers and Bunce, 1991). The co-planar chlorinated biphenyls are mostly substituted in the meta and/or para positions. Mono- and di-ortho substituted PCBs undergo turning of their phenyl rings and subsequent loss of flexibility and co-planar structure. Di-ortho PCBs belong to the PB-type of inducers. One therefore speaks of mixed-type inducers with regard to PCBs (van Straalen and Verkleij, 1991).

According to Nebert et al. (2000), more recent findings demonstrate that the 3MC-type compounds, such as TCDD binding the AhR, activate a battery of genes (called Ah battery). The Ah battery consists of at least two P450 genes (CYP1A1 and CYP1A2) and four non-P450 genes expressing enzymes involved in phase 2 biotransfromation reactions and antioxidant activity. The promoters of the P450 genes of the Ah battery contain an AhRE (XRE), whereas the promoters of the non-P450 genes of the Ah battery contain both an AhRE (XRE) and an electrophile response element, EpRE, (ARE). This implies that TCDD activates all genes of the Ah battery via the AhREs, whereas oxidative stress can activate the non-P450 genes of the Ah battery via the EpREs. However, some compounds of the 3MC-type inducers, such as benzo(a)pyrene (BaP), that are readily metabolized to reactive intermediates, activate genes of the Ah battery via both the AhREs and the EpREs. It is not unlikely that the reactive intermediates of phase 1 biotransformation by P450 enzymes would then follow an oxidative stress pathway leading to activation of EpREs of the Ah battery genes. This could be accomplished by activation of a mitogen-activated protein kinase (MAPK) route by decreased glutathione (GSH) levels.

Moreover, TCDD, although a 3MC-type compound, also stimulates the activation of the DNA-binding factor AP-1 (Jun/Fos). Puga et al. (1992) demonstrated that TCDD brings about this effect by activating phospholipase C (PLC) and PKC and releasing Ca^{2+}. TCDD also induces the expression of c-jun and c-fos genes, resulting in a large increase in AP-1 concentration. This leads to the speculation that TCDD could also activate genes of the Ah battery by binding of AP-1 to the EpREs on the promoters of the Ah battery genes. AP-1 would be activated by PKC, which, in turn, would be activated by PLC and Ca^{2+} (see Figure 2.1). This detailed analysis demonstrates that the activities of the 3MC-type and PB-type compounds are not so strictly separated as supposed previously. It means that differentiation of expression of P450 and related genes is caused by integration of combined effects. It also means that xenobiotics and oxidants can induce genes belonging to the Ah battery without being ligated to the AhR. The analysis is an example of MFO activity and oxidative stress response. Figure 6.2 summarizes the various pathways leading to the induction of P450 enzymes and related compounds.

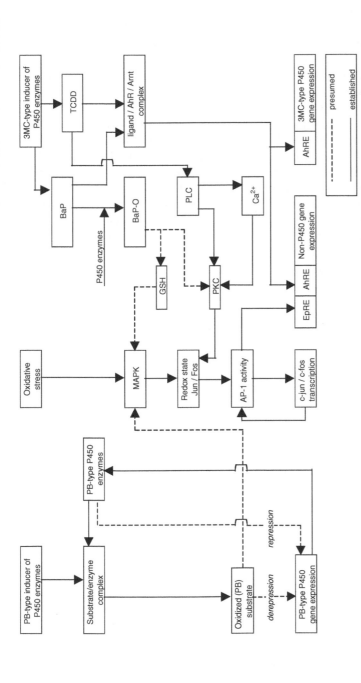

Figure 6.2 Summary of the various pathways leading to the induction of P450 enzymes and related compounds belonging to the Ah battery of genes. PB-type compounds probably induce P450 enzymes by substrate–enzyme binding; 3MC-type compounds ligate the AhR complex, which then activates genes of the Ah battery via AhREs on the promoters of Ah battery genes. Alternatively, 3MC-type compounds could activate AP-1, either directly via PKC (e.g., by TCDD) or indirectly via the MAPK route (e.g., by BaP). This speculative route is shown by dashed lines. Other oxidants not activated by biotransformation could follow the same route to activate genes of the Ah battery. Abbreviations: PB-type = phenobarbital-type; 3MC-type = 3-methylcholanthrene-type; MAPK = mitogen-activated protein kinase; AP-1 = activator protein-1; BaP = benzo(a)pyrene; BaP-O = oxidized BaP; PKC = protein kinase C; PLC = phospholipase C; TCDD = tetrachlorodibenzo-p-dioxin; AhR = aryl hydrocarbon receptor; Arnt = Ah-receptor nuclear translocator; AhRE = Ah regulatory element; EpRE = electrophile response element. (Activation of Ah-battery genes has been partly adapted from Nebert, D.W. et al., *Biochem. Pharmacol.* 59: 65–85, 2000.)

Hydrophobic endo- and xenobiotics are converted into hydrophilic compounds by a variety of cytochrome P450 enzymes during phase 1 biotransformation. Depending on the type of substrate and cellular microenvironment (aerobic, low oxygen tension, anaerobic), the enzymes exhibit a mono-oxygenase, an oxidase, or a substrate reductase activity (Goeptar, 1993). At these different forms of reaction, the cytochrome P450 enzymes are supported by nicotinamide adenine dinucleotide phosphate hydride (NADPH)–cytochrome P450 reductases, which donate the required electrons for these reactions. These activities are explained in Figure 6.3.

Figure 6.3 shows the different activities exhibited by cytochrome P450 enzymes and their interrelationships. The main activity under normal aerobic conditions is the mono-oxygenase activity, of which hydroxylation (SH + O_2 + NADPH + H^+ → SOH + H_2O + $NADP^+$, S = substrate) is frequently occurring. Thus, one oxygen molecule is cleaved, at which point one atom contributes to hydroxylation and the other forms water. The required two electrons are donated by NADPH and transferred via cytochrome P450 reductase and cytochrome P450. Depending on the type of substrate, either enzymes of the P450 IA class (3MC-type substrates) or of the P450 IIB class (PB-type substrates) catalyze the hydroxylation step (Timbrell, 1991).

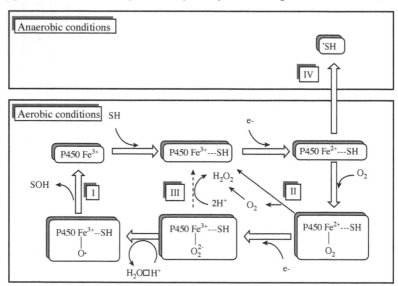

Figure 6.3 The reaction cycle of cytochrome P450. Cytochrome P450 is conveniently indicated as P450 Fe^{3+} and the substrate, as SH. Upon binding of the substrate to the catalytic site of P450, the P450-substrate complex is reduced in a one-electron reduction by NADPH-cytochrome P450 reductase. Subsequent binding of molecular oxygen followed by a second electron transfer results in substrate mono-oxygenation (pathway I). Uncoupling reactions (leading to oxidase activity) compete with substrate oxygenation (pathways II and III). Substrates can also be directly reduced by ferrous P450 under anaerobic or low-oxygen tension conditions (pathway IV). (After Goeptar, A.R., The role of cytochrome P450 in the reductive bioactivation of cytostatic quinones: a molecular toxicological study. Ph.D. thesis, Vrije Universiteit, Amsterdam, 1993.)

The P450 reaction cycle can be short-circuited in such a way that O_2 is reduced to either $O_2^{\bullet-}$ or H_2O_2, instead of being used for substrate oxygenation (pathways II and III of Figure 6.3). This compromising side reaction is often referred to as "uncoupling" or "oxidase activity" of P450. Goeptar (1993) also introduced the term "oxygen reductase activity" for these oxygen reduction reactions. This activity is stimulated by specific substrates in relation to specific enzymes. For instance, 2,3,5,6-tetramethylbenzoquinone (TMQ) stimulates the oxygen reductase activity in the presence of the cytochrome P450 IA1 enzyme, whereas hexobarbital stimulates the same reaction in the presence of the cytochrome P450 IIB1 enzyme (Goeptar, 1993).

Under anaerobic conditions, P450 enzymes demonstrate their substrate reductase activity by transferring electrons to their substrates in a one-electron reduction step. Quinonoid compounds in particular are of interest for this study, because, in a cellular environment of low-oxygen partial pressure, the formed semiquinones may be oxidized again, starting a redox cycling process under the formation of superoxide anion radical ($O_2^{\bullet-}$) (Cadenas, 1994; Goeptar, 1993; Timbrell, 1991). This process takes place in the transition phase between aerobic and anaerobic conditions, where low oxygen tension prevails. The initiation of such a redox-cycling process further depends on the type of compound, its redox potential, and the cellular redox state. One remarkable fact observed is that TMQ is reduced under anaerobic conditions by PB-type P450s, such as P450 IIB1 (Goeptar, 1993). This seems to be in contrast with what has been mentioned above about TMQ. It demonstrates the complexity of the interactions of P450 enzymes and that the events that commit P450 to reduce substrates are not yet fully understood (Goeptar, 1993).

6.3 The mixed-function oxygenase system in insects

Cytochrome P450 enzymes in insects are found in several important tissues, such as the digestive tract, the Malpighian tubules, and the fat body. It is expected that they are also expressed in specialized tissues, such as antennae, and in "first line of defense" tissues, such as the midgut (Brun et al., 1996; Feyereisen, 1999). They fulfill, as in higher animal species, many important tasks, from the synthesis and degradation of ecdysteroid and juvenile hormones, to the metabolism of foreign chemicals of natural or synthetic origin. This diversity in function is achieved by the insect genome, probably carrying about one hundred P450 genes (Feyereisen, 1999).

The knowledge of insect P450 is largely based on assumptions about homology to the well-studied mammalian P450 systems. To date, insect P450s have been assigned to six cytochrome P450 (CYP) families (CYP nomenclature according to Nebert et al., 1987). Five are insect specific (CYP 6, 9, 12, 18, and 28); CYP 4 includes sequences from vertebrates (Adams et al., 2000; Feyereisen, 1999). All have different tasks and are differently expressed. For instance, some P450s are larva specific (e.g., CYP 6B2) whereas others are adult specific (e.g., CYP 6D1) (Feyereisen, 1999; Scott et al., 1998). CYP 6A2 has a wide distribution in *D. melanogaster,* and its genes (Cyp 6a2) are under control of the molting hormone, 20-hydroxyecdysone (Feyereisen, 1999; Spiegelman et al., 1997).

Although little is known of P450 gene expression in insects, it is likely that the same pattern of expression as in higher animal species is followed, because identical pathways, enzymes, and transcription factors exist in these species. For instance, the 3MC-type and PB-type pathways for induction of specific P450 enzymes in mammalian species are also reported for insects (Fuchs et al., 1994). Pohjanvirta and Tuomisto (1994) reported on the existence of the Ah receptor in *D. melanogaster.* Spiegelman et al. (1997) demonstrated the presence of the AHH enzyme in this species. Surprisingly, they reported on the induction of this P450 enzyme by PB, whereas this enzyme would be expected to be induced by inducers of the 3MC-type.

In general, the biochemistry of biotransformation phases 1 and 2 resembles the description of the MFO system in Section 6.1. The classes of P450 enzymes in insects differ from those in vertebrates, however. In phase 2, other conjugating compounds, such as phosphates and glucose, are used in insects (Lafont and Connat, 1989).

6.4   Degradation of xenobiotics in insects

During phase 1 biotransformation, P450 enzymes exhibit mono-oxygenase, oxidase, or substrate reductase activity. At these biochemical processes, harmful intermediates or reactive oxygen species may be formed, and they may inhibit P450 enzymes (Feyereisen, 1999), attack cellular macromolecules, or activate other cellular defense systems. This aspect will be further discussed in Section 9.1.3.4 because it emphasizes the role of the MFO system in relation to other cellular defense systems.

One notable example, in this respect, of substrate reductase activity is the one-electron reduction of quinonoid compounds to semiquinones by P450 enzymes. Many of them are quickly autoxidizable with the concomitant production of superoxide anion radical, giving rise to redox cycling and potential cytotoxicity (Cadenas, 1994; Chaudère, 1994; Goeptar, 1993; Timbrell, 1991) (illustrated in Figure 6.4).

Figure 6.4 shows the involvement of cytochrome P450 enzymes of the MFO system in the metabolism of quinonoid compounds. By means of a one-electron reduction, the quinone (Q) is converted into the semiquinone ($SQ^{\bullet-}$), which may be regenerated to the original quinone structure by autoxidation, depending on the type of quinone and the redox state. During this autoxidation process, superoxide anion radical is formed. A subsequent redox cycling can be minimized if the semiquinone species is readily converted to quinone and hydroquinone ($2SQ^{\bullet-} + 2H^+ \rightarrow QH_2 + Q$) (Cadenas, 1994). The involvement of the MFO system and possible $O_2^{\bullet-}$ generation can also be reduced by binding of quinone to GSH or by an enzymatic two-electron reduction of the quinone species to hydroquinone. In *Drosophila,* this reduction is effected by the enzyme diaphorase-1 (DIA-1), the equivalent of

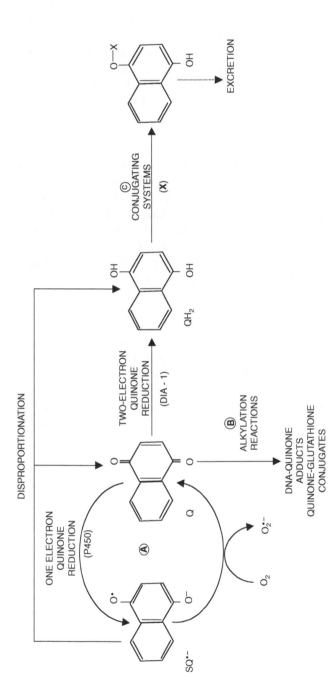

Figure 6.4 Production of superoxide anion radical by the one-electron reduction of quinonoid compounds and autoxidation of the semi-quinone (SQ•⁻), regenerating the original quinone structure (Q). The reduction step involves cytochrome P450 enzymes and NADPH-cytochrome P450 reductases. This route A also shows the alternative to autoxidation, that is, the conversion of the semiquinone to quinone and hydroquinone (QH₂). Route B shows interactions of quinone with cellular compounds, whereas route C shows the removal of quinones by a two-electron reduction forming hydroquinone as a first step. In *Drosophila*, the enzyme diaphorase-1 (DIA-1) is involved in this step. The second part of this route includes conjugation. In insects, X stands for phosphates or glucose. (Adapted from Cadenas, E., in *Free Radicals in the Environment, Medicine and Toxicology: Critical Aspects and Current Highlights*, Nohl, H. et al., Eds., Richelieu Press, London, 1994.)

DT-diaphorase in mammalian species (Georgieva et al., 1995). A fast metabolism of quinone species is desirable, since these species and their intermediates may exert other harmful effects as well. They may, for instance, form adducts with DNA (Cadenas, 1994; Cerutti, 1985).

Insects are confronted more specifically, however, with insecticides, and herbivorous species, with plant toxicants as well. Two important groups of synthetic insecticides are organophosphorous compounds, which inhibit acetylcholinesterase activity (Feyereisen, 1999; Timbrell, 1991), and pyrethroids, which disturb sodium/potassium transport by damaging cellular membranes (van Straalen and Verkleij, 1991).

Organophosphorous compounds may undergo a bioactivation step (P = S to P = O) by mono-oxygenase activity of P450 enzymes or become detoxified by ester cleavage (Feyereisen, 1999). Pyrethroids induce the PB-type P450 enzyme CYP6D1, in many insects, including house flies. By mediating mono-oxygenase activity, CYP6D1 provides resistance to these pyrethroids (Kasai and Scott, 2000; Liu and Scott, 1998; Scott et al., 1998). The aspects of resistance to insecticides will be discussed further in Chapter 7.

In summary, P450 enzymes of the MFO system play an active role in downgrading insecticides and other xenobiotics, such as quinonoid compounds, PAHs, PCDDs, and PCBs. In this way, the MFO system makes a positive contribution to cellular homeostasis. However, bioactivation of P450 substrates and production of reactive oxygen species may take place, which has a negative effect on cellular defense and burdens other cellular defense systems. If these negative side effects cannot be eliminated by other defense systems, they may cause damage to cellular structures and compromise performance, which may lead to cell death. To prevent these effects, other cellular defense systems provide a cooperative response, and the MFO system can therefore be considered a member of the cellular defense system.

chapter 7

The role of cellular response systems in developing tolerance to environmental stress

7.1 Introduction

If cells or organisms are exposed to environmental stress, it is possible that they will develop or demonstrate tolerance. This enhances the chance of survival and extends longevity. Tolerance may have a genetic basis in adaptation through natural selection operating on genetically based interindividual variation in tolerance. This type of tolerance is inheritable. Tolerance may also be acquired, however, by physiological adjustments in response to pre-exposure to sublethal concentrations of the stressor. This acclimatory tolerance may be induced and lost within a generation and cannot be passed on to offspring (Posthuma and van Straalen, 1993). This acclimation is considered a type of phenotypic plasticity, which is defined as the extent to which environmental variation can modify the expression of a genotype at the phenotypic level (Hoffmann and Parsons, 1991).

The acclimatory process leading to tolerance can be divided into long-term tolerance persisting for, for example, a season and short-term tolerance lasting, for example, a few days. For the acquisition of tolerance to seasonal fluctuations, the term *acclimatization* is also used. For describing the acquisition of a short-term tolerance, the term *acclimation* is often used. Thermotolerance is described by Hahn and Li (1990) as the acquisition of a transiently increased resistance to heat or cold. Both heat and cold stress, and the acquisition of tolerance to these stresses will be discussed in the next section. The term *thermoresistance* is related to inheritable tolerance and has a more permanent character. In this chapter, focusing on thermo- and chemotolerance and their effects on life-history traits in relation to the induction of cellular stress defense systems, the term *tolerance* will be used in the sense of acquired, acclimatory tolerance. Inheritable tolerance, at least the genetic analysis of this type of tolerance, will only be discussed if necessary. It will then be explicitly mentioned as such.

The acquisition of tolerance by the induction of cellular stress defense systems may be accomplished by different mechanisms. In the case of heat stress, the cell has to defend itself against a physical stress, first causing denaturation of proteins. To resist this denaturation process, the cell induces heat-shock proteins (HSPs) that protect essential proteins, promote refolding, or repair damage. In the case of oxidative stress, oxidants are neutralized by scavengers of the antioxidant system and detoxified and prepared for excretion by enzymes of the mixed-function oxygenase system. Metal tolerance is achieved by binding of heavy metals to metallothionein (MT). This is considered a matter of sequestration, because the metal ion cannot be degraded. Another form of sequestration is compartmentalization, which may have a temporary or permanent basis, with temporary storage being followed by excretion. This enumeration of tolerances to specific stressors demonstrates that each type of stressor evokes cellular tolerance by specific mechanisms. In the following sections, the acquisition of tolerance by various terrestrial arthropod species, with emphasis on *Drosophila,* and in relation to specific cellular defense mechanisms, will be discussed. Tolerance has its benefits with regard to life-history traits, such as survival and longevity, but may also involve costs, and related trade-offs will therefore be discussed as well.

7.2　Thermotolerance in insect species

7.2.1　Tolerance to heat

Acquired (acclimatory) tolerance to heat can be divided into tolerance to short-term thermal fluctuations, for example, over several hours, or long-term, such as seasonal, fluctuations. Both types of acquired tolerance are temporary and reversible, based on physiological adjustments. Within this framework, the description of short-term tolerance used by Hahn and Li (1990) fits well; that is, heat tolerance is the acquisition of a transiently increased resistance to heat. It appears to be a universal phenomenon, observed in all organisms tested. Little is known, however, of the mechanisms that lead to this type of tolerance, resulting in shorter recovery times and enhanced survival of heat-tolerant cells after heat stress (Black and Subjeck, 1990). It is known, however, that tolerance to heat correlates with accumulation of HSPs (Bensaude et al. 1996). This will be further analyzed below.

Tolerance to heat can be developed by exposing cells to a sublethal heat-shock treatment, followed by a recovery period, also called a recovery interval, at normal physiological temperatures. During this recovery process, the cell builds up tolerance, which results in increased cell survival after a subsequent and otherwise lethal heat-shock treatment (Mizzen and Welch,

1988). For instance, in mammalian cells, tolerance is induced during the exposure of the cells to 41 to 42.5°C. At these temperatures, "normal" protein synthesis is only partially inhibited, whereas at temperatures reaching 44 to 45°C, normal protein synthesis is completely inhibited (Black and Subjeck, 1990). At recovery at 37°C after conditioning, which may last, for example, for 8 hours, tolerance is built up, while at the same time normal protein synthesis is resumed. Tolerance may then persist for at least 48 hours, and may even last longer, depending on the severity of the conditioning heat shock (Dahlgaard et al., 1998). Heat-tolerant cells are more resistant not only to heat but also to other stressors, as shown in survival tests. Moreover, with regard to the resumption of normal protein synthesis, heat-tolerant cells show shorter recovery times than nontolerant cells after exposure to severe stress (Black and Subjeck, 1990). However, tolerance can also be developed by continuous heating, whereby protein synthesis continues (Hahn and Li, 1990). It is not clear how this level of tolerance compares with the tolerance evoked by heat-shock treatment.

Assuming that the ambient temperature for *Drosophila* is 25°C, whereas their larvae in such a climate may undergo temperatures up to 44.5°C in necrotic fruit (Feder, 1996), the temperature range in which heat tolerance is developed may be as large as 10 to 15°C. Pauli et al. (1992) reported on tests in which *Drosophila* were exposed to a severe heat shock (about 40°C), and the majority of the animals died. When, however, just prior to the severe stress, the animals underwent a mild heat shock (about 33°C), sufficient to activate the synthesis of HSPs during the interval period without disturbing normal protein synthesis, a significant increase in survival was observed. It was supposed that the HSPs induced during the mild heat shock and interval period protect the organism from lethal stress and thus contribute to heat tolerance. It was further suggested that several different types of the HSP family are involved in the protection exhibited by heat tolerance. Krebs and Loeschcke (1994) tested the optimal interval time between conditioning and severe heat stress in relation to maximum tolerance, measured in survival tests. They found that survival was optimal if the flies were conditioned 8 hours before exposure to severe stress, but that survival was reduced if conditioning took place only 2 to 4 hours before exposure to severe stress. By 8 hours of recovery, synthesis of HSPs of the sp70 family was completed. Moreover, conditioning and defense against severe stress cost energy. Thus, if the interval time is too short, the cell may still have an energy debt when exposed to severe stress (Krebs and Loeschcke, 1994).

Because little was known at the time of the thermal ecology of *Drosophila*, Feder et al. (1997) simulated field conditions for larvae and pupae in necrotic fruit and examined the levels of hsp70. In the laboratory, temperatures around 40°C kill *Drosophila* larvae rapidly, and no good explanation could be given for *Drosophila* larvae resisting higher temperatures in the field. A gradual transfer from 25°C to the test temperatures resulted in no less mortality than did direct transfer. Moreover, at gradual heating, only a modest increase of the hsp70 level was observed. These tests demonstrate that there is still little understanding of the mechanisms underlying heat tolerance, but they provide an ecological context for further studies of heat stress and tolerance in *Drosophila*.

Arrigo (1998) reported on the role of small HSPs (SHSPs) in resistance to heat shock and their contribution to heat tolerance. Cells overexpressing SHSPs revealed an enhanced resistance to further heat shock. Multiple mechanisms are believed to underlie this resistance. First, SHSPs would accelerate the recovery from the heat-induced shut-off of RNA processing and protein synthesis. Second, overexpression of SHSPs would counteract the disruption of the actin microfilament network. Third, large oligomers of SHSP would interact with nonnative proteins and create a reservoir of folding intermediates that accelerate a restart of the folding process after heat stress, by other chaperones, such as hsp70.

According to Parsell and Lindquist (1994), hsp70 appears to be the major protein involved in *Drosophila* tolerance to extreme temperatures (40 to 42°C). This was tested by expressing hsp70 from a heterologous promoter at 25°C, without the induction of other HSPs. It appeared to be sufficient to provide tolerance to direct exposure to 42°C. However, at normal temperatures, the increased hsp70 levels appeared to be detrimental to growth and cell division. Dahlgaard et al. (1998) examined the functional relationship between hsp70 and heat tolerance in adult *Drosophila melanogaster* and concluded that hsp70 may be an important contributor to heat tolerance but that this tolerance probably involves a cascade of actions by HSPs. They base this conclusion on the observation that after the hsp70 level had fallen to the base level, 32 hours after build-up of tolerance, resistance measured in survival tests still remained elevated. Loeschcke et al. (1997) observed the same effect. The increase in survival after development of tolerance persisted much longer than hsp70 expression. These results also suggest that other factors than, or in addition to, hsp70 concentration are responsible for the variation in heat tolerance in *Drosophila*.

With regard to RNA processing, Black and Subjeck (1990) emphasize that among RNAs, ribosomal RNA (rRNA) is most sensitive to heat, and its processing is inhibited first. Realizing that HSPs are supposed to protect rRNA, Black and Subjeck (1990) consider rRNA protection and subsequent prolonged protein synthesis to be important factors in heat tolerance. Indeed, Welch and Mizzen (1988) observed that the perturbation of the small nuclear ribonucleoprotein (snRNP) complexes after heat-shock treatment was less apparent in tolerant cells than in nontolerant cells exposed to a subsequent heat-shock treatment. This conclusion is related to the observation by Yost et al. (1990) that RNA splicing was inhibited in *Drosophila* cells after a severe heat-shock treatment. Together, these observations could explain why heat-tolerant cells demonstrate a prolonged synthesis of normal proteins compared with nontolerant cells. Referring to effects on intermediate filaments, the heat-induced collapse of the cytoskeleton did not occur in cells rendered heat tolerant. A third aspect, observed by Welch and Mizzen (1988), was that the redistribution of both constitutive and induced HSPs of the sp70 family from the cytoplasm and nucleus into the nucleolus was faster in heat-tolerant cells than in nontolerant cells. One might predict that this results in a faster return to normal nucleolar functioning in heat-tolerant cells than in nontolerant cells. Mizzen and Welch (1988) further observed that heat-tolerant cells exhibited considerably less translational inhibition and an overall reduction in the amount of subsequent stress protein synthesis. They hypothesize that the underlying mechanism for the observed differences between a heat-tolerant and a nontolerant cell is the faster kinetics of both the synthesis and repression of the stress proteins and their accelerated redistribution after a second and more severe heat-shock treatment.

On the basis of findings by Laszlo (1992), Bensaude et al. (1996) mentioned two mechanisms underlying heat tolerance in relation to increased HSP levels: accelerated damage repair and protection against initial damage. Repair of damaged proteins by HSP is a widely accepted phenomenon and is supported by the capability of chaperones to refold denatured proteins. Protection against initial damage is demonstrated by overexpression of hsp70. Both effects lead to an attenuated aggregation and inactivation of intranuclear proteins (Stege et al., 1994). These conclusions do not counter the findings by Welch and Mizzen (1988), and they all subscribe to the important role of HSPs in heat tolerance.

Thus, there is a correlation between HSP kinetics and heat tolerance, but this is independent of the origin of the stress. For instance, other exogenous stressors, such as ethanol, also induce HSPs and thus heat tolerance (Black and Subjeck, 1990). The converse is also true: Heat shock induces tolerance to ethanol, anoxia, and some other forms of stress (Lindquist, 1986). However, induction of HSPs by amino acid analogs does not lead to heat tolerance, because aberrant HSPs are formed. Hence, it can be concluded that functional HSPs are required for developing heat tolerance (Black and Subjeck, 1990).

The question remains of whether HSPs are more heat resistant than other proteins and, if so, why? McKay et al. (1994) reported that bound MgATP/ADP stabilizes the ATPase region of hsp70 in bovine cells against thermal unfolding. Considering the chaperoning function of HSPs (see Ellis and van der Vies, 1991), it might also be possible that HSPs and interacting proteins stabilize each other against denaturation and formation of insoluble aggregates by their interaction (Krebs and Loeschcke, 1994). Bensaude et al. (1996) reported that prevention of protein aggregation by chaperones might occur only above a specific temperature threshold. As an example, they mentioned that α-crystallin, which is homologous to a C-terminal domain in SHSPs, exposes hydrophobic surfaces at temperatures above 30°C and prevents aggregation of certain molecules above that temperature. An increasing efficiency in preventing aggregation is obtained at higher temperatures up to 41°C. We speculate that together, heat-shock protection and subsequent heat tolerance is accomplished by the interaction of denaturing proteins and (induced) heat-shock proteins. Neutral sequences exposed by unfolding proteins interact with hydrophobic sites of HSPs by means of van der Waals attractive forces and hydrogen bonds (Hightower et al., 1994). In this way, inappropriate interactions between proteins are prevented, while at the same time renaturation after the heat shock is made possible by the weak interactions. At this renaturation process, the HSPs assist in a correct stepwise refolding, as postulated by Hartl (1996). According to Feder et al. (1992), excess HSPs could also lead to the formation of insoluble aggregates (for further explanation, see below). This observation supports our view that denaturation of both "normal" proteins and their protectors against heat shock, the HSPs, is prevented by their interaction, and that they mutually benefit from their way of bonding.

Krebs and Feder (1997a) investigated differential expression of hsp70 in various tissues of larvae of *D. melanogaster*. An immunofluorescence technique was used to compare hsp70 expression, whereas tissue damage was indicated by staining with trypan blue. Brain, salivary glands, imaginal disks, and hindgut expressed hsp70 within the first hour of heat shock, whereas gut tissues, fat body, and Malpighian tubules did not express hsp70 until 4 to 21 hours after heat shock. These tissues stained most intensively. Tissues responding fast to heat shock by expressing hsp70 lost these stress proteins soon at recovery. The tissues slowly expressing hsp70 after heat shock extended their expression during recovery. It may therefore be supposed that these HSPs are induced as a response to damage caused by heat shock rather than to elevated temperature. These aspects relate to the method of activation of heat-shock factor (HSF), that is, indirect or direct, respectively (see Chapter 3). Coincidentally, those tissues showing rapid induction of hsp70 all have a common embryonic origin in the ectoderm. Some gut tissues (endoderm tissue) showed intensive staining, suggesting that their function does not obtain priority in acute survival. This tissue-specific expression seems to have a genetic basis and it is an example of how selective evolution has been in providing the ability of expressing hsp70 during heat stress. The fact that its rate and/or level of expression is limited in certain tissues not only determines heat tolerance for the whole organism but also means that expression of hsp70 has substantial negative effects on the cell, even under stressful conditions.

Feder et al. (1992) also investigated the consequences of expressing hsp70 in *Drosophila* cells at normal temperatures, using heterologous promoters to force hsp70 expression. Initially the rate of cell growth was reduced substantially, but with continued expression, cell growth rates recovered. At the same time, hsp70 molecules coalesced into discrete hsp70 granules. The protein in these granules appeared to be irreversibly inactivated. It cannot be dispersed with a second heat shock, and cells containing these granules do not show heat tolerance. The cell seems to control the protein's activity by sequestration. Feder et al. (1992) consider expression of hsp70 at normal temperatures detrimental to growth. Indeed, hsp70, the most inducible protein type of the sp70 family in *Drosophila*, is hardly detectable under normal conditions. On the contrary, hsp72, the inducible heat-shock protein of the sp70 family in human cells, is present in measurable concentrations, together with its constitutive cognate, hsp73. This means that in *Drosophila*, the inducible hsp70 and the constitutively present hsc70 are different, notwithstanding the fact that their functions, certainly under stressful conditions, may overlap.

To test the effects on heat tolerance of overexpressing hsp70, Feder et al. (1996) engineered extra-copy larvae and pupae of *D. melanogaster*. They established that hsp70 was undetectable prior to heat shock. Hsp70 concentrations were higher in the extra-copy strain than in the control group during and after heat shock. Heat tolerance significantly improved in the extra-copy strain. The experimental conditions resembled thermal regimes actually experienced by *Drosophila* in the field. However, newly hatched larvae of the extra-copy strain died at greater rates than the control larvae when their HSP levels were increased continuously by repeated mild heat shock. Possibly, in such circumstances, hsp70 may bind target proteins inappropriately. This observation suggests that the wild-type hsp70 gene copy number and levels of hsp70 expression may have evolved as a compromise between the beneficial and deleterious effects of hsp70 (Feder et al., 1996). To investigate this aspect further, Krebs and Feder (1997b) extended the tests and observed that the extra-copy larvae strain surpassed the control strain in survival only immediately after heat stress. Control larvae survived to adulthood at higher proportions than did extra-copy larvae and grew more rapidly after heat stress. It was therefore concluded that, although extra hsp70 provides additional protection against the immediate damage from heat stress, abnormally high concentrations can decrease growth, development, and survival to adulthood.

Krebs and Feder (1997c) investigated the ecological relevance of these findings by looking at natural variation in the expression of hsp70 in a population of *D. melanogaster* and its correlation with heat tolerance. They characterized variation in first-instar larvae of 20 isofemale lines isolated from a single natural population of *D. melanogaster*. They found that hsp70 expression varied more than twofold among lines after induction by exposure to 36°C for 1 hour. As expected, expression of hsp70 correlated positively with larval heat tolerance, but hsp70 expression was inversely correlated with survival in the absence of stress. The capability of inducing hsp70 seems to be an evolutionary trade-off between heat tolerance under stressful conditions and survival in the absence of stress (Krebs and Feder, 1997c).

Loeschcke et al. (1997) investigated the costs and benefits of tolerance at the level of populations, using iso-female lines of *Drosophila buzzatii* and *D. melanogaster*. One remarkable observation was that for populations with relatively high survival rates at one life stage, survival often was low at another life stage. Furthermore, survival and fecundity of acclimated and nonacclimated flies were compared, and decreased fertility following increased survival after acclimation was identified as a cost of expressing the heat-shock response. Loeschcke et al. (1997) observed, however, that the duration of reduced fecundity was similar to the duration of the heat tolerance and that this reduced fecundity therefore was also a reversible acclimatory response. According to Krebs and Loeschcke (1994), the costs of activating heat-shock proteins are not large, especially compared with the costs associated with stress in their absence, which makes the widespread presence of the heat-shock response across taxa not surprising.

Another striking example of costs and benefits of increased expression of a cellular defense system is given by Kuether and Arking (1999). They found that *Drosophila* flies with normal life spans exhibit a heat-induced longevity extension, but that long-lived animals already resistant to oxidative stress exhibit a heat-induced longevity shortening. The underlying mechanisms are not known, but Arking (1998) showed that there appears to exist an inverse relationship between the levels of antioxidant defense enzymes and the levels of enzymes specific for other types of stresses, such as cytochrome P450 enzymes, as well as between the antioxidant defense enzymes and heat-shock proteins. Extended longevity from an increased resistance to oxidative stress in the adult *Drosophila* fly may result in a decreased fitness to other important environmental parameters, such as heat tolerance.

Optimization seems not only applicable among different stress defense systems, but also within one system. Krebs and Feder (1998) observed that, while small to moderate increases in hsp70 levels enhance inducible heat tolerance in *Drosophila*, large increases in hsp70 levels actually decrease heat tolerance. Evolution thus may favor an intermediate level of hsp70 (Feder and Hofmann, 1999). One reason that extended longevity does not appear to be common in the wild may be the inability of such organisms to thrive in changeable environments characterized by multiple stressors (Kuether and Arking, 1999). These findings seem to conflict with the hypothesis of the integrated cellular stress response system. One should realize, however, that the above-mentioned relationships relate to fitness during life, whereas an integration of stress response relates to an acute phase, with the purpose of surviving at any cost. That does not alter the fact that availability of amino acids may determine levels of induction.

Seasonal fluctuations in temperature take place over periods of months. For fluctuations of such length, the types of short-term adjustments discussed above would be inadequate. Accordingly, organisms can supplement these short-term adjustments with others that reversibly alter their sensitivity to longer-term fluctuations. Such physiological reorganizations are also called acclimatization responses. The mechanisms of acclimatization are highly diverse, including synthesis of different isozymes, differential utilization of metabolic pathways, altered rates of protein synthesis and degradation, and changes in membrane composition (Huey and Bennett, 1990). It is not clear whether the heat stress response shows similar seasonal fluctuations. If so, Huey and Bennett (1990) suppose that the threshold temperature for the induction of HSPs will be lower in winter than in summer. Feder and Hofmann (1999) postulate that, in general, the threshold temperature for HSP induction is correlated with the typical temperatures at which species live, with thermophilic species having a higher threshold than psychrophilic species.

Apart from that, it is questionable whether HSPs fulfill specific roles in relation to seasonal fluctuations (Feder and Hofmann, 1999). It is not likely that something would change with regard to the inducible hsp70 in *Drosophila*, because its presence has deleterious effects in the absence of stress. Furthermore, there is no information available concerning the variation in basal levels of constitutively present heat-shock proteins during winter and summer. Basal levels and inducibility of HSPs, and related threshold temperatures, may vary, however, between *Drosophila* species more permanently living in different habitats. Huey and Bennett (1990) reported on a comparative study of heat tolerance and heat stress response in larvae of *D. melanogaster* and two desert species of *Drosophila*. The larvae were raised at 25°C and acutely exposed to a graded series of temperatures. The results are summarized in Table 7.1.

Table 7.1 shows a shift toward higher temperatures for maximum hsp70 synthesis for the desert species. This may relate to a shift in threshold temperature as well. An identical pattern is shown for temperatures of maximum tolerance. Such data suggest that protein synthesis may have an upper thermal maximum that varies among species adapted to different temperature environments (Feder and Hofmann, 1999). In this respect, it should be realized that these differences have a permanent character and may arise from genetic variation and that this type of tolerance may be qualified as inheritable tolerance or heat resistance.

7.2.2   Tolerance to cold

Overwintering insects in temperate and cold climates have to cope with long periods of freezing conditions. These ectotherm animals have to protect themselves by specific measures to survive these severe conditions. For that purpose, they have developed two alternative strategies: tolerating or avoiding freezing. A summary of these strategies is given below. For further reading about this subject, see Lee and Denlinger (1991).

Table 7.1 Comparison of heat stress response and heat tolerance limits of *Drosophila melanogaster* and two desert species of *Drosophila*

Species	Habitat	Temperature at maximum hsp70 synthesis	Shutdown of all protein synthesis
D. melanogaster	Widespread	35 to 37°C	37°C
D. mojavensis	Desert	39°C	41°C
D. arizonensis	Desert	41°C	43°C

Source: After Huey, R.B. and Bennett, A.F., in *Stress Proteins in Biology and Medicine*, Morimoto, R.I. et al., Eds., Cold Spring Harbor Laboratory Press, Cold Spring Harbor, New York, 1990. With permission.)

Freeze-tolerant animals accept freezing of their extracellular fluid (hemolymph). During the freezing process, an osmotic pressure is built up by a concentration of solutes in the remaining fluid. This leads to cellular dehydration and subsequent prevention of intracellular freezing and membrane damage. Organisms may even promote this process by producing "ice nucleating agents" (INAs), which may initiate freezing by acting as nuclei for the growth of ice crystals. In fact, INAs increase the supercooling point (SCP), the temperature at which spontaneous nucleation of body fluids occurs, thereby reducing freezing and dehydration rates and preventing cellular damage. These INAs may be active in the hemolymph of animals.

Freeze-intolerant (susceptible) species avoid freezing of any compartment and therefore try to lower their actual freezing point. To minimize their risk of unexpected nucleation, they remove endogenous nucleating substances by metabolism and sequestration in membrane lipids. This effect lowers the SCP of the animal.

To further regulate their range of freezing in relation to their environmental conditions, both freeze-tolerant and freeze-intolerant insects may produce two different groups of compounds:

1. Low-molecular-weight (LMW) carbohydrates consisting of polyols, such as glycerol and sorbitol, and sugars, such as glucose, fructose, and trehalose
2. Antifreeze proteins (AFPs), such as glycoproteins, rich in alanine, threonine, and disaccharides derived from galactose

LMW carbohydrates (also called cryoprotectants) decrease the melting point and the SCP of the hemolymph in insects. The melting point is decreased in relation to the concentration of the solutes and their osmolality (1.86°C/osm) (Zachariassen, 1991). Furthermore, it has been suggested that this type of antifreeze decreases the SCP by masking or inactivating ice nucleators (Lee, 1991). This role seems to be illogical in freeze-tolerant species. It has to be realized, however, that, for example, glycerol, the most common cryoprotectant in insects, fulfills a variety of other functions, such as:

1. Protecting insects against injury caused by cold shock (discussed below)
2. Preventing excessive dehydration and membrane damage at a too-high extracellular osmolality, by diffusion of glycerol into the intracellular environment (Chen et al., 1987; Lee, 1991; Zachariassen, 1991)
3. Preventing the redistribution and segregation of membrane components upon thawing, when the cell is in a hypertonic state (Chen et al., 1987; Lee, 1991; Zachariassen, 1991)

AFPs are also referred to as "thermal hysteresis proteins," because they produce a thermal differential between the hemolymph melting and freezing points. In freeze-intolerant species, they lower both the hemolymph freezing point and the SCP, the latter by adsorption of AFP to embryonic ice crystals (Lee, 1991). The AFPs are not osmotically active, which has the advantage that the overwintering insect can remain active under these conditions (Lavy, 1996). The role of these AFPs in freeze-tolerant species seems to be the inhibition of recrystallization of ice crystals within frozen tissues (Lee, 1991).

This description of compounds and their various functions illustrates the complex regulation of both freeze-tolerant and freeze-intolerant species to achieve optimal conditions for overwintering and survival, in relation to their ecological conditions. On average, freeze-tolerant species have SCPs in the range of −10°C, whereas freeze-intolerant species strive for SCPs below −20°C in temperate climates. This cold-hardening process associated with overwintering takes a few weeks. Short-term cold hardening, which may take place during chilling periods in spring and autumn, is a much faster response. An insufficient response may lead to a cold shock, causing direct injury to membranes owing to phase transition in membrane lipids, resulting in leakage. Cold shock may also cause thermoelastic stress and membrane rupture, owing to the membrane condensing to a greater extent than the contents of the cell (Lee, 1991). These types of injuries are observed in the absence of ice formation at temperatures well above the SCP (Denlinger et al., 1991). Protection against these types of injuries are provided by rapid cold-hardening (RCH), after which the production of cryoprotectants may be observed within an hour. It is different from winter cold-hardening, which may occur in an inactive or diapausing stage of insects; RCH may occur in feeding and even reproductively active insects (Czajka and Lee, 1990).

D. melanogaster is freeze intolerant and is susceptible to cold shock (Czajka and Lee, 1990; Misener et al., 2001). To survive winter conditions, this species develops an SCP of −20°C (Czajka and Lee, 1990). From their tests, Czajka and Lee concluded that larvae, pupae, and adults of the fruit fly are all able to overwinter. Whether *Drosophila* enter a state of diapause depends on the species and the overwintering conditions. For instance, within *Drosophila triauraria* there are both diapausing and nondiapausing strains depending on the area (Kimura et al., 1992). According to Czajka and Lee (1990), *D. melanogaster* does not appear to overwinter in diapause, but rather seems to pass the winter months in a state of quiescence. They further observed that *D. melanogaster* flies die when exposed to −5°C for 1 hour. However, when first chilled at 0°C for 1 hour, most survive this otherwise lethal cold exposure. This RCH effect was seen in larvae, pupae, and adults of this species. Similar tests were executed by other investigators, who confirmed the RCH effect in *Drosophila* (Chen and Walker, 1994; Misener et al., 2001). Kelty and Lee (1999, 2001) repeated these tests at ecologically relevant cooling rates (0.1°C/min) and observed a significantly higher survival after 1 hour exposure to −7°C than for those cooled at 1°C/min. Kelty and Lee (1999) also observed that the critical thermal minimum (i.e., the temperature at which torpor occurred) dropped from 6.5 to 3.9°C when cooled at ecologically relevant rates.

According to Kelty and Lee (1999), the mechanisms by which *D. melanogaster* builds up tolerance remain elusive. They were unable to detect any whole-body glycerol in their tests. In their more recent tests, Kelty and Lee (2001) found that no cryoprotectants of a carbohydrate nature, including glycerol, were produced during an RCH treatment. Nor did they find the production of hsp70 during their cold-conditioning tests. On the contrary, Burton et al. (1988) already established the induction of HSPs upon a cold-shock treatment of *D. melanogaster*, but it appeared that these HSPs were actually induced during recovery. Tsutsayeva and Sevryukova (2001) hypothesize that some hsp70 is induced after cold shock to repair proteins damaged by cold shock. However, the recovery from cold shock can also be considered a kind of heat shock in view of a temperaure difference of about 25°C. Burton et al. (1988) emphasize that protein denaturation, which forms the basis for the induction of HSPs, is very unlikely to occur as a result of cold treatment. On the contrary, a mild heat shock prior to a cold treatment enhanced the tolerance to cold shock. Whether the induction of HSPs or some other compound provided the protection is not clear. It is known, however, that some species induce polyols at high temperature (Chen et al., 1987). Kelty and Lee (2001) also observed the induction of HSPs upon return from cold treatment.

D. melanogaster does not make use of cryoprotectants or HSPs for protection against cold shock. AFPs may contribute to lowering the melting point and SCP of *D. melanogaster*, but that could be independent of the phenomenon of cold-shock injury, such as membrane damage. Some investigators suggest that changes in membrane composition take place to protect against cold-shock injury in both vertebrate and invertebrate species. These changes result in maintenance of membrane fluidity over a wider range of temperatures than would otherwise be the case. Such homeoviscous adjustments occur during both long-term and short-term acclimation (Kelty and Lee, 2001; Ohtsu et al., 1998). The degree of unsaturated lipids would be higher in cold-tolerant species than in cold-susceptible species, but Ohtsu et al. (1998) observed contradictory results in *D. melanogaster* species from temperate and subtropical areas. They noticed that with enhancement of cold tolerance, the number of double bonds in the phospolipid decreased without a marked variation in the percentages of unsaturated fatty acids. Whether this leads to a lower membrane fluidity in *D. melanogaster* is not clear, considering that phosphatidylethanolamine is a dominant phospholipid in this species, while many other species have phosphatidylcholine-rich membranes (Ohtsu et al., 1998).

Chen and Walker (1994) emphasize the importance of energy storage as a factor in enhanced cold tolerance. Selection for tolerance to cold shock or chilling over several generations of *D. melanogaster* resulted in a significant increase in tolerance to these cold treatments. Higher contents of glycogen and total protein were observed in the cold-selected lines than in the control line. With respect to the importance of energy storage in relation to cold tolerance, it is worthwile mentioning that Misener et al. (2001) found a higher proline content in *D. melanogaster* lines selected for resistance to chilling injury. This higher content may contribute to the energy pool associated with overwintering, but being an osmolyte, proline could also fulfill a role in influencing the freezing process.

In late autumn, when the temperature falls, *Drosophila* species induce the sugar trehalose and maintain high levels of it during winter at the expense of glycogen (Kimura et al., 1992). The species in the colder areas maintained higher levels of glycogen. This shows that stored energy in the form of glycogen is also a factor in developing tolerance to survival during winter. The same applies for triacylglycerols. The fact that triacylglycerol levels are two to five times higher in diapausing than in nondiapausing species supports this statement (Kimura et al., 1992). The function of trehalose as a cryoprotectant is, first, lowering the melting point and the SCP, but as an energy carrier, it also contributes to the energy pool (Kimura et al., 1992; Ohtsu et al., 1998). Apart from this function, trehalose plays a role in *Drosophila* in other stress responses, for example, in tolerance to anoxia. This protection is likely to be due to a reduction of protein aggregation (Chen et al., 2002).

It may be concluded that a good picture of how *Drosophila* protect themselves against short- and long-term winter conditions has not yet been established. In several aspects of cold conditioning, as described in the introductory part of this section, *Drosophila* differs from other insect species. It may also be concluded that the induction of heat shock proteins is not an issue in cold hardening and certainly not in overwintering.

7.3   Chemotolerance in insect species

Terrestrial arthropods are often confronted with a variety of environmental pollutants, such as polycyclic aromatic hydrocarbons (PAHs), pesticides, and heavy metals. This may be due to, for example, soil pollution or it may be on purpose, for example, by application of insecticides. At the cellular level, arthropods may defend themselves by activating specific cellular defense systems. In the case of chemical stress, the oxidative stress response system, the mixed function oxygenase (MFO) system, and the MT system are important to defending the cell by neutralizing, degrading, or sequestering the penetrated pollutant. By means of induction, tolerance may be built up and then slowly disappear again after diminution of the stress below a certain threshold. In some cases, by means of mutation and selection, resistance to the stress is demonstrated by maintaining base levels of stress-defense enzymes and proteins. This resistance has a permanent character because of inheritability of the trait. However, it is often not clear which of the two types of tolerance is observed, and inheritable tolerance on the basis of genetic differences has only been demonstrated in a few cases.

In this section, two subjects related to chemotolerance will be discussed: tolerance to insecticides and tolerance to heavy metals. In the former case, the dominating role of the MFO system is illustrated, in the latter, MTs play an important role.

7.3.1   The role of the mixed-function oxygenase system in tolerance to insecticides in insects

Insects are confronted with insecticides, and herbivorous species with plant toxicants as well. Two important groups of synthetic insecticides are organophosphorous compounds, which inhibit acetylcholinesterase activity (Feyereisen, 1999; Timbrell, 1991), and pyrethroids, which disturb sodium/potassium transport by damaging cellular membranes (van Straalen and Verkleij, 1991).

Organophosphorous compounds may undergo a bioactivation step (P = S to P = O) by the mono-oxygenase activity of cytochrome P450 enzymes or become detoxified by ester cleavage. The bioactivation of the organophosphorous compounds results in substantially increased activity as an inhibitor of acetylcholinesterase (Feyereisen, 1999). Resistance to these insecticides has been observed in several insect species as well. This resistance has been linked to insensitive acetylcholinesterases and sodium channels (Fogleman et al., 1998), but it is also related to overproduction of esterase B2, as a result of gene amplification. This phenomenon has been observed in *Culex* mosquitoes and has now become a worldwide resistance mechanism (Kasai and Scott, 2000; Raymond et al., 1991). For more detailed information about cytochrome P450 enzymes as part of the MFO system in insects, see Chapter 6.

Pyrethroids may induce various types of cytochrome P450 enzymes and subsequently be metabolized by mono-oxygenase-mediated activity of these enzymes (Kasai and Scott, 2000). For instance, in the housefly, *Musca domestica*, the cytochrome P450 enzyme CYP6D1 seems to be responsible for providing resistance to pyrethroids (Feyereisen, 1999; Kasai and Scott, 2000; Liu and Scott, 1998; Scott et al., 1998; Tomita et al., 1995), whereas in *Helicoverpa armigera* this role is taken by CYP6B2 (Ranasinghe et al., 1997) and CYP6B7 (Ranasinghe and Hobbs, 1999). The type of P450 enzyme may be life-stage dependent, however. In the case of *H. armigera*, whose larvae take in monoterpenes by feeding on plants of the mint family, one could speak of tolerance. These terpenes induce CYP6B2, especially in the midgut and fat body, and build up elevated levels of these enzymes during the larval stage (Ranasinghe et al., 1997). However, CYP6B2 and CYP6B7 also metabolize pyrethroids and thus provide tolerance to this insecticide indirectly (Ranasinghe et al., 1997; Ranasinghe and Hobbs, 1999). The induction of CYP6B7 only occurred in fat bodies. Ranasinghe and Hobbs (1999) also observed that feeding fifth-instar larvae with the pyrethroid permethrin, resulted in a direct but slight induction of CYP6B7. This was due to concentrations of permethrin being too low to prevent a substantial knock-down effect.

CYP6D1 seems to be responsible for resistance to pyrethroids in the housefly, *M. domestica*. This resistance is not developed by amplification of the gene expressing CYP6D1 (Fogleman et al., 1998) but by a constitutive overexpression of this enzyme, controlled by transacting regulatory factors

(Feyereisen, 1999). This has been demonstrated in a housefly strain called "learn pyrethroid resistant" (LPR) strain, which was collected in New York State. In this strain, CYP6D1 is expressed ninefold higher (mRNA and protein) than in susceptible strains (Liu and Scott, 1998). In contrast to the induced resistance to pyrethroids by enzymes of the CYP6B family, this example of pyrethroid resistance by CYP6D1 is a form of inheritable resistance. To determine whether CYP6D1 is involved in mono-oxygenase-mediated pyrethroid resistance in other populations, Kasai and Scott (2000) compared two housefly strains from Georgia, one collected in 1983 and the other in 1998. The most recent strain showed identical characteristics of resistance to pyrethroids as the LPR strain from New York State. Thus, this trait has spread, but how common it is worldwide is not known.

Alternatively, cytochrome P450 enzyme inhibitors of synthetic origin, such as methylenedioxyphenyl (MDP), and of natural origin, such as sesamin and sesamolin, both MDP-type inhibitors, may synergize pyrethroid activity and toxicity by inhibition of cytochrome P450 enzymes (Feyereisen, 1999). Oxidative metabolism of the methylene group of MDP leads to the formation of a carbene group that forms a virtually irreversible complex with the heme iron of cytochrome P450 enzymes. This is a form of mechanism-based "suicide" inhibition (Feyereisen, 1999).

Drosophila also demonstrate chemotolerance in insect–host plant relationships. A well-known example described by Fogleman et al. (1998) is the cactus–*Drosophila* model system of the Sonoran desert in the southwestern United States. The relationship involves five species of allelo-chemical-containing columnar cactus species and four species of endemic cactophilic drosophilids. An almost one-to-one relationship exists between the *Drosophila* and cactus species. Each *Drosophila* species has independently specialized in the chemically unique decomposing tissue of a single cactus species, and cytochrome P450s are involved in this host–plant specialization. Primary allelo-chemicals in these cactus species include alkaloids and medium-chain fatty acids. The former group of chemicals may block steroid metabolism or nerve cell function; the latter may inhibit oxidative phosphorylation. It is not clear whether the specialization of the drosophilids is based on genetic variation. Feyereisen (1999) reported that in three of these cactophilic species, the response of 7 to 11 cytochrome P450s to inducers has been studied and four classes of induction could be observed. It supports his view that, in general, a subset of widely divergent cytochrome P450 genes is recruited by an inducer from the repertoire of cytochrome P450 genes and that the subsets of P450 enzymes induced by different classes of inducers overlap and form a grid of induction pathways. It should be mentioned, however, that of the 26 responses among the three *Drosophila* species only three cytochrome P450 genes were strongly induced. Indeed, Fogleman et al. (1998) does not exclude the possibility that a few (possibly only one) substrate-specific cytochrome P450s are responsible for the bulk of cactus allelo-chemical detoxification in each of the Sonoran drosophilids.

7.3.2 The role of metallothionein in tolerance to heavy metals in Drosophila

Drosophila are confronted with heavy metals in the environment, predominantly by uptake of food in the digestive tract. Heavy metals are absorbed in special metal-specific tissues of the fly. For instance, after severe exposure, copper is mainly absorbed in cuprophilic cells of the midgut, zinc and copper in the Malpighian tubules, and cadmium in midgut epithelium cells (Durliat et al., 1995). Epithelium cells in the midgut take up heavy metals both from the lumenal side and the hemolymph (Dallinger, 1993; Posthuma and van Straalen, 1993). Cellular absorption of these heavy metals may in the first place lead to the induction of MT, which binds to these metals and prevents toxic effects. In the midgut, copper and cadmium are mainly bound to the MTN type of *Drosophila* MT, whereas zinc in the Malpighian tubules, although a weak inducer, is mainly bound to the MTO type (Durliat et al., 1995). These observations do not fully comply with the observations made by Bonneton et al. (1996) and referred to in Chapter 5. For a more detailed description of the induction of MT, different types and functions of MT, and their localizations in the cell, see Chapter 5.

By binding to MT, heavy metals are inactivated and cellular toxicity is prevented. Since the metal itself is nondegradable, this form of detoxification could be classified as sequestration. The midgut epithelium cells are also able to sequester heavy metals by compartmentalization, that is, storage in membrane-enclosed vesicles (granules) as part of the lysosomal fraction in the cell, where each type of metal is bound irreversibly to a specific compound in the granule (Dallinger, 1993; Hopkin, 1989; Posthuma and van Straalen, 1993). Hopkin (1989) describes four types of granules, which have also been observed in midgut epithelium cells of *Drosophila* (Lauverjat et al., 1989). By compartmentalization, excess amounts of heavy metals are stored. Some insect species are able to eliminate the metal by cellular excretion processes, such as exocytosis of vesicles (Dallinger, 1993; Lauverjat et al., 1989). Thus, in these species, accumulation of metals is limited.

Drosophila also possess the ability to excrete vesicles (Lauverjat et al., 1989) and can therefore be considered intermediate species with regard to heavy metal accumulation. The sequestration capacity depends on the number of granules and the capacity of MT as a storage compound as well as a carrier of heavy metals into the lysomal fraction of the cell (Posthuma and van Straalen, 1993). Other species, such as snails and isopods, seem not to possess such an excretion mechanism, and metal accumulation is therefore unavoidable. These species are therefore considered macroconcentrators (Dallinger, 1993). However, in collembolan species, such as the springtail *Orchesella cincta*, the old midgut epithelium is shed during molting, and accumulated heavy metal is eliminated at the same time (Posthuma, 1992). These animals are therefore considered deconcentrators of heavy metals (Dallinger, 1993).

D. *melanogaster* is the first species in which tolerance to heavy metals was demonstrated. Maroni et al. (1987) found that natural populations of this species differed in amplification of the MT gene. Strains with amplified MT genes had a higher resistance to copper and cadmium. One of the two MT genes in *D. melanogaster*, Mtn, occurs in two allelic forms, Mtn[1] and Mtn[3]. Maroni et al. (1995) reported that Mtn[1] was found at frequencies of 85% and 95% in American and European samples, respectively. Mtn[3] was the minority allele in these samples but was fixed in a sample from Congo. Moreover, duplication of the Mtn[1] allele has been observed in natural and laboratory populations, and the strains carrying duplicated Mtn[1] alleles tolerate increased metal concentrations (Bonneton et al., 1996). Maroni et al. (1995) assume that the duplications of Mtn[1] in European species are a response to the agricultural use of copper salts for antimicrobial purposes. In this respect, it is worth mentioning that MTN[1] has a lysosomal localization, whereas MTN[3] is in the cytosol. It may be speculated that increased tolerance by Mtn[1] duplication correlates to this localization and sequestration of copper and cadmium. Together, this increased tolerance has a genetic basis and may be a good example of how *Drosophila* has adapted to the enhanced metal pollution in the industrialized world. This adaptation seems to be twofold: dominance and duplication of the Mtn[1] gene.

Shirley and Sibly (1999) reported on development of cadmium tolerance in a natural *D. melanogaster* population maintained in the laboratory for 20 generations. Increased cadmium tolerance was reflected by enhanced juvenile survivorship, shorter developmental period, and increased fecundity compared with the control population. However, resistant lines paid a fitness cost in unpolluted environments, with fecundity and emergence weights being reduced. Two mechanisms have been proposed that might lead to such fitness costs. First, a detoxification mechanism might use energy and other resources. Second, resistant individuals may be less efficient at metal uptake, leading to micronutrient deficiency in unpolluted environments. The genetic basis for the evolved life-history differences was investigated with crosses and backcrosses between the lines. The results of this study suggest that the evolved cadmium resistance was due to a single, dominant, sex-linked gene on the X-chromosome (Shirley and Sibly, 1999). Gill et al. (1989) had already suggested that one or more genes on the X-chromosome control the expression of structural genes for MT. If, indeed, only one gene is involved, then the gene has pleiotropic effects on fecundity, survivorship, and adult weight, via the expression of MT (Shirley and Sibly, 1999). This study was based on artificial selection, acting upon genetic variation in the gene for metal resistance. The result is a "between-environment trade-off," with large antagonistic, pleiotropic effects, allowing animals to increase fitness in polluted environments but only at the cost of reduced growth and reproduction in unpolluted environments (Shirley and Sibly, 1999).

Callaghan and Denny (2002) continued the work of Shirley and Sibly (1999) by investigating the relationship between p-glycoprotein (pGP) and cadmium toxicity. pGP is an ATP-dependent transporter, which pumps molecules out of cells and is responsible for multidrug resistance. Previous work by Broeks et al. (1996) showed that when the pGP gene in cadmium-resistant *Caenorhabditis* nematodes was inactivated, the worms became sensitive to cadmium toxicity. Callaghan and Denny (2002) used the wild-type and cadmium-resistant *Drosophila* strains of Shirley and Sibly (1999) and tested the effects of adding different concentrations of verapamil, a calcium-channel blocker, to the larval diet, both in the presence or absence of 80 mg/kg cadmium chloride. Blocking or partly blocking pGP increased the toxicity of cadmium to the *Drosophila* larvae, reducing the number of adults emerging. This suggests that pGPs have a role in the removal of cadmium from the cell to reduce toxicity. Because the effect of verapamil was smaller on the cadmium-resistant strain than on the wild-type strain, it was questioned whether the difference in effect could be ascribed to over-expression of pGP by gene duplication. These tests further raise the question of to what extent transporters are involved in cell/tissue specificity with regard to defense against environmental stressors.

In summary, these examples of metal tolerance in *D. melanogaster* give an impression of the various pathways in these species and, in general, in most terrestrial invertebrates, leading to either acclimatory or inheritable tolerance. However, the results are often not clear enough to allow for a sharp differentiation between these two extreme forms of tolerance. One often has to deal with a combination of the genetic and physiologic component. What has been discussed in this section is summarized in Figure 7.1.

Figure 7.1 shows a physiologic and a genetic component, leading to acquired and inheritable tolerance, respectively. The physiologic pathway comprises sequestration of heavy metals by binding to induced MT and by compartmentalization or excretion of heavy metals. The genetic pathway includes gene amplification and selection on gene variation. All these processes may provide the animal with tolerance to heavy metal pollution in its environment. The examples of inheritable tolerance described above show that, according to evolutionary theories, adaptation leads to increased fitness but only in relation to the circumstances to which the animal has adapted.

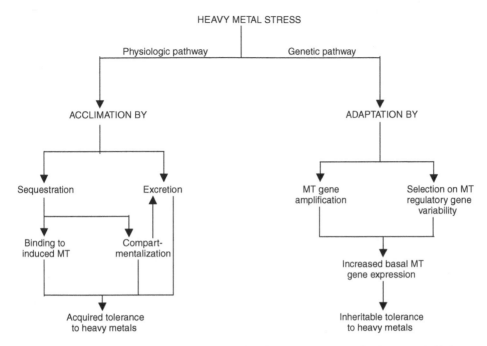

Figure 7.1 Various pathways in *Drosophila* that lead to heavy metal tolerance. Cellular stress by heavy metals may lead to a physiological response and subsequent, acquired tolerance or a genetic response and subsequent, inheritable tolerance. The physiologic pathway comprises sequestration or excretion of heavy metals. Sequestration is achieved by binding to induced metallothionein and by compartmentalization. The genetic pathway includes gene amplification and selection on gene variation.

HEAVY METAL STRESS

AURELIAN SMES

chapter 8

The effects of environmental stress on aging and cell death

8.1 Introduction

Environmental stressors exhibit their adverse effects at different levels of organization, such as populations, individual organisms, tissues, and cells. At the individual level, these effects are demonstrated in the life-history traits of an organism, such as development, growth, aging, longevity, survival, and reproduction. However, whatever the level of exhibition of the damage or the trait affected is, the basic attack and subsequent physiological changes take place in cells of organisms. To counteract these effects and minimize damage, a cell possesses various stress defense systems, described in previous chapters of this book.

This chapter will describe the processes of aging and cell death. One aim will be to translate the process of aging at the cellular level to the organism level. The main focus in this review will be on the role of cellular stress defense systems in both processes. The scope of this review is mainly related to terrestrial arthropods. Since most of the research on the phylum of the Arthropoda is related to *Drosophila* species, emphasis will be on these species. The process of aging will be analyzed first, followed by the process leading to cell death.

8.2 Effects of environmental stress on aging

8.2.1 Introduction

The term *aging* is primarily related to senescence of the mature organism, finally leading to its death and determining its longevity. In this context, gerontologists have long investigated this complex subject and observed the striking correlation between species-specific metabolic rate and life span: Species with higher metabolic rates consume more O_2 per gram of tissue and have a shorter maximum life-span potential (MLSP). As a result, the life energy potential (LEP), defined as the product of MLSP and specific metabolic rate, is roughly a constant for virtually all species. Thus, in 1928 it was concluded that energy consumption was responsible for senescence, a concept referred to as the "rate of living hypothesis" (Beckmann and Ames, 1997; Kleiber, 1961). However, this hypothesis is based on a descriptive phenomenology of aging in organisms without offering molecular insights (Yu, 1994).

Investigations of the production of reactive oxygen species (ROS), such as $O_2^{\bullet-}$, $OH^{\bullet}$, and H_2O_2, and protein oxidation in the mitochondria of flies as a marker for longevity and aging, led to the conclusion that these factors increase with age and are inversely correlated with life-span potential (Beckmann and Ames, 1997; Richter and Schweizer, 1997). Orr and Sohal (1994) reported on tests with transgenic *Drosophila* overexpressing the antioxidant enzymes superoxide dismutase (SOD) and catalase (CAT). These flies exhibited a one-third longer life span, a delayed loss of physical performance, and a decrease in oxidative damage of protein and DNA compared with control flies. These results suggest that the main factor relating oxidative stress to aging and longevity is not oxygen consumption but the rate of ROS production (Guyton et al., 1997; Richter and Schweizer, 1997). These observations lead to the "free radical theory of aging," which purports that free radical reactions, arising largely in the course of normal metabolism, are the underlying deleterious factors responsible for the aging process (Beckmann and Ames, 1997; Yu, 1994). This theory offers both molecular and mechanistic explanations for the complex aging phenomenon (Yu, 1994). Free radicals may damage major cellular constituents, such as nucleic acids, lipids, and proteins, which may lead to an impairment of cellular physiological functions that are involved in the degradation or regeneration of damaged molecules.

Evolutionary biologists have significantly contributed to the free radical theory of aging by explaining why physiologically harmful generation of oxygen free radicals occurs. They have argued that natural selection favors genes that increase reproduction, rather than genes that act to preserve nongerm cells (the "disposable soma"). In short, selective pressure to survive and compete effectively at an early age may work against the conservation of the soma in the long run (Beckmann and Ames, 1997). This "disposable soma theory of aging" supports the idea that aging is caused by the accumulation of random molecular defects in the cell (Parsons, 1995, 1996). This support is based on the observation that production of reactive oxygen species and subsequent oxidation of molecules is inherent to aerobic respiration. Kowald and Kirkwood (1994) argued for a "network theory of aging," integrating the free radical theory of aging and other aging theories, so that an underlying molecular and physiological basis for aging is emerging (Parsons, 1995, 1996).

The production of ROS depends on the balance between oxidation rate and antioxidant capacity. Sources of ROS production are mainly found in the generation of superoxide anion radicals by either endogenous factors, such as respiration and inflammatory response, or exogenous factors, such as environmental stressors. These environmental stressors may therefore contribute to additional ROS production and subsequently to the aging process, in line with the free radical theory of aging and, more generally, with the network theory of aging. An analysis of the aging process, the contribution of environmental stressors, and the involvement of cellular stress defense systems will be the subject of this section.

8.2.2   The aging process under normal and stressful conditions

Aging is considered a continuous process, which proceeds during an organism's life, including developmental stages, such as metamorphosis of insects. It culminates in the death of the organism. As mentioned in the Introduction, the term aging is primarily related to the senescence of the whole organism, but the underlying biochemical reactions and the physiological effects occur at the cellular level. These effects result in age-related phenotypic expressions at the level of the organism. Therefore, it would be better to speak of deterioration of cells and aging of organisms, so this terminology will be followed here.

According to Oliver et al. (1987) and Lundberg et al. (2000), deterioration of cells seems to be independent of the cell passage number (number of times the cell reproduces itself by division), at least over the intermediate range of passages. This is based on tests with cultured human fibroblasts, showing that the amount of oxidized protein varies with the age of the fibroblast owner. This suggests that genetic changes may be responsible for these manifestations (Stadtman, 1992). Deterioration of cells is also different for various cell types and tissues: Relatively unspecialized cells that continue to divide during life do not age as rapidly as specialized cells that have lost the capacity to divide. Within the body of a multicellular animal, tissues such as muscle and nerve, in which cell division has normally ceased, slowly deteriorate, whereas tissues such as those of the liver and pancreas, in which active cell division continues, age much more slowly. Therefore, the aging and subsequent death of an organism is related to the deterioration and death of cells and tissues that cannot be replaced (Keeton and Gould, 1993). Tests performed by Arking (1998) with *Drosophila* strains showed that the developmental stages do not have a direct relationship to adult longevity.

Furthermore, the rate of aging of an organism depends on its physiological age, not on its chronological age. Thus, insects with a short life span age faster than mammals with a much longer life span. Assuming that the free radical theory of aging and the disposable soma theory underlie the aging process, then the rate of aging is determined by the following three major factors (Guyton et al., 1997):

1. The rate of production of reactive oxygen species
2. The scavenging and detoxification capacity of antioxidant systems
3. The capacity to repair oxidative lesions and remove damaged molecules

8.2.2.1   Rate of production of reactive oxygen species

The superoxide anion radical ($O_2^{\bullet-}$) and hydrogen peroxide (H_2O_2) are the major components of the family of ROS. Four important sources contribute to the generation of $O_2^{\bullet-}$: the respiratory chain in mitochondria, cytochrome P450 enzymes in the microsomal fraction of the endoplasmic reticulum, the xanthine oxidase pathway, and redox cycling processes. H_2O_2 is predominantly produced by dismutation of $O_2^{\bullet-}$ to H_2O_2, catalyzed by the enzyme superoxide dismutase (SOD). Other enzymatic reactions generating H_2O_2 are the two-electron reduction of O_2 to H_2O_2 by monoamine oxidase (MAO) and the peroxisomal β-oxidation of fatty acids. Peroxisomal H_2O_2 is normally decomposed by the enzyme CAT (Ames et al., 1993; Beckmann and Ames, 1997).

8.2.2.2   Scavenging and detoxification capacity of antioxidant systems

The major scavenger of oxidants is glutathione (GSH), which not only scavenges ROS, but also influences to a great extent the redox state by regulating the balance of reduced and oxidized thiols. Other important scavengers are vitamins A, C, and E, which are mainly involved in the repression of lipid peroxidation. Detoxification of ROS is mainly performed by the enzyme systems SOD, CAT, and glutathione peroxidase (GSH-Px).

8.2.2.3   Capacity to repair oxidative lesions and remove damaged molecules

ROS may cause lipid peroxidation, oxidation of sulfhydryl groups in proteins, and damage to DNA. The capacity to repair oxidative lesions and remove damaged molecules thus implies repair and degradation systems for proteins, lipids, and nucleic acids. An example of an important degradation system is the proteolytic system in the cytosol of the cell. Protein, which cannot be repaired, will be degraded in this system. Stress proteins, such as hsp70 and ubiquitin, are involved in this system. This degradation pathway has been described in Chapter 3.

According to Guyton et al. (1997), the rate of production of ROS increases during life, whereas the scavenging and detoxifying capacity as well as the repair and degradation capacity for damaged molecules decrease. For instance, Sörensen and Loeschcke (2002) demonstrated that the inducibility of the important stress protein hsp70 in *Drosophila melanogaster* is reduced during aging. Thus, it is likely that the accumulation of damaged molecules during life is progressive. On the basis of the aging theories mentioned above, it is believed that the accumulation of unrepaired damage could account for the age-related loss of physiological functions (Stadtman, 1992). Since ROS exert their oxidizing effects not only to DNA but also to lipids by lipid peroxidation and to proteins by oxidizing sulfhydryl groups, accumulation of damaged molecules occurs throughout the cell. Consequently, damaged membranes may not only affect gene expression but also the functioning of membranes and enzymes in a direct way. The free radical theory of aging was further refined by incorporation of the concept of the "somatic mutation theory of aging," which states that the accumulation of DNA mutations is responsible for degenerative senescence (Beckmann and Ames, 1997). The effects on DNA contribute, together with effects from damaged lipids and proteins, to impairment of the physiological functioning of the cell and to aging of the organism.

Since multiple processes are acting in parallel, leading to numerous aging theories, Kowald and Kirkwood (1994) argued for a "network theory of aging" and developed as a first step a mathematical model integrating the ideas of the free radical theory of aging and the "protein error theory." The latter describes the process of error propagation in protein synthesis. The model describes aspects of random errors, radicals, antioxidants, and scavengers and is therefore called the ERAS model. It is a step forward in explaining experimental observations.

The explanatory power of the model stems from the integration of different biochemical processes. The model supports the observation that oxidative stress increases with age. It shows a decrease in antioxidant activity that resembles experimental observations, such as the decrease in activity of SOD and CAT with age observed in house flies. The activity declines because too many of the newly synthesized proteins are inactive or erroneous (Kowald and Kirkwood, 1994). The same effect is observed with the activity of proteolytic enzyme systems. Consequently, a decline in turnover rate, an increase in the half-life of proteins, but a reduction in the level of correct proteins is observed (Kowald and Kirkwood, 1994).

Orr and Sohal (1994) tested the free radical theory of aging and by extension the network theory of aging by overexpressing cytosolic copper and zinc SOD and CAT in *D. melanogaster*. The transgenic flies exhibited as much as a one third extension of life span, a lower amount of oxidative damage to proteins, and a delayed loss of physical performance. The results provided direct support for the free radical theory of aging. The investigators realized that an increased life span without a corresponding increase in metabolic potential would not be a meaningful gain from the perspective of the aging process. Therefore, they measured metabolic rate as well and concluded that the metabolic potential of the transgenic flies was higher than that of the controls. To determine if the overexpression of SOD and CAT genes lowered the *in vivo* oxidative damage to the tissues, a comparison of protein carbonyl content (a measure of oxidized protein level) was made between the control and the transgenic flies. At all ages, the protein carbonyl content of transgenic flies was notably lower than that of the controls (Orr and Sohal, 1994). In more recent but identical studies performed by Sohal et al. (1995), the activity of glucose-6-phosphate dehydrogenase (G6PDH), an indicator of oxidative damage, was determined. The relatively slower age-related loss of the activity of this enzyme, observed in the transgenic flies, reflect the correspondingly lower levels of oxidative damage. Similarly, the oxidative damage to DNA was determined by measuring the accumulation of 8-hydroxydeoxyguanosine (8-OHdG) during life. The slope of the age-related increase in 8-OHdG accumulation was lower for the transgenic flies than for the controls. These experiments also support the concept that ROS production is a causal factor in the aging process. Arking (2001), Barja (2002), Sohal (2002), and Sohal et al. (2002) reconfirmed this opinion, but they all emphasize that a direct link between oxidative stress and aging has not yet been established and that additional evidence is needed to clearly define the nature of the involvement. Moreover, aging is also likely to be a multifactorial process and not reducible to any single cause (Pletcher et al., 2002; Wickens, 2001). For further details about this aspect, see the discussion of the genetic compound of aging, below.

Mockett et al. (1999a) examined the role of mitochondria in the aging process by overexpressing manganese SOD (MnSOD) in a range of 5 to 116% in 15 experimental lines of *D. melanogaster*. The results indicated that the mean longevity of the experimental lines was decreased by 4 to 5% relative to the controls. The naturally evolved level of MnSOD activity in *Drosophila* appears to be near the optimum required under normal conditions, although the optimum may be shifted to a higher level under more stressful conditions. McCord (1995) is of the opinion that an optimum level of $O_2^{\bullet-}$ exists, and decreasing amounts of $O_2^{\bullet-}$ below this level will actually increase the overall level of oxidative stress. A mechanistic basis for this observation is the finding that $O_2^{\bullet-}$ can both initiate and terminate lipid peroxidation. Thus, very low levels of $O_2^{\bullet-}$ may result in excessive chain propagation and low rates of termination. Mockett et al. (2003) further examined the role of mitochondria in the aging of *Drosophila* species by ectopic expression of CAT in these organelles. The result was that stress resistance was increased but longevity was not. From these and the other tests, it may be concluded that antioxidant defense capacity in mitochondria seems to be near the optimum with regard to aging. Mockett et al. (1999b) also tested transgenic *D. melanogaster* flies overexpressing glutathione reductase under both hyperoxia and normoxia. Under normoxic conditions, overexpression of glutathione reductase had no effect on longevity, protein carbonyl content, glutathione, or glutathione disulfide content. Mockett et al. (1999b) concluded that below a certain threshold rate of pro-oxidant production, as in normal aging, the accumulation of oxidative damage is relatively insensitive to changes in antioxidant levels, possibly because these oxidants react with specific targets in their immediate vicinity.

It may be concluded that aging is related to a continuous degeneration of cells during an organism's life and that the reduction of cellular performance is probably a consequence of three major factors:

1. Unavoidable random damage of (macro)molecules during life under homeostatic conditions
2. Genetic components controlling enzymatic activities by changes in gene expression

3. Exogenous factors, such as environmental stressors, usually amplifying the aging process

8.2.3 Major factors in cellular degeneration

8.2.3.1 Unavoidable damage

The production of ROS, especially during respiration in aerobic cells, is an unavoidable consequence of aerobic life. Referring to the evolutionary theory of aging, Beckmann and Ames (1997) state that life is about trade-offs, and it appears that a certain amount of O_2 toxicity is written into the evolutionary contract. Increased ROS production with age, owing to a decline in the activity of the respiratory chain in mitochondria, among other factors, has been observed (Shigenaga et al., 1994). This will be discussed below. As discussed above, the resulting oxidative damage leads to accumulation of aberrant molecules, which is believed to be a major contributor to aging.

Destruction of biological molecules by oxidants primarily involves lipid peroxidation. A primary effect of lipid peroxidation is decreased membrane fluidity, which alters membrane properties and can considerably disrupt membrane-bound proteins (Shi et al., 1994; Yu, 1994). This is the case not only in plasma membranes but also in mitochondrial membranes. Because of the relationship with increased ROS production in mitochondria, this is now discussed in more detail. Mitochondrial membrane fluidity is not only affected by lipid peroxidation. It has been observed in rodents that, during life, there is a progressive change in fatty acid composition of the phosphoglyceride cardiolipin toward a higher degree of unsaturation, making the mitochondrial membrane more susceptible to oxidation (the peroxidizability index increases). Cardiolipin plays a pivotal role in facilitating the activities of mitochondrial inner-membrane enzymes. An example is the reduction of the activity of the adenosine triphosphate/adenosine diphosphate (ATP/ADP) translocator, leading to reduced respiration by a shortage of ADP and an increased production of $O_2^{\bullet-}$ and H_2O_2, causing oxidative damage (Shigenaga et al., 1994). Another consequence of unsaturation of cardiolipin is an increased permeability of the inner membrane, resulting in a loss of proton gradient. These effects and many others lead to age-related histological changes, including mitochondrial enlargement, matrix vacuolization, and shortened cristae (Shigenaga et al., 1994). Thus, mitochondrial decay is indicative of cellular degeneration and is an important contributor to aging.

The oxidation of protein is less well characterized, but several classes of damage have been documented, including oxidation of sulfhydryl groups, oxidative adduction of amino acid residues via metal-catalyzed oxidation, reaction with aldehydes, protein-protein cross-linking, and peptide fragmentation (Beckmann and Ames, 1997). Important aspects of protein oxidation are the loss of enzyme activity and various membrane functions. With regard to aging, Stadtman (1992) reported that, while in young individuals, up to about 10% of the total protein content is damaged, in aged individuals, up to 30% of the proteins, measured as carbonyls, were in an oxidized form. In view of the highly sophisticated regulation of enzyme activity, substantial decreases in the amounts of enzymes and the accumulation of massive amounts of damaged proteins, as occurs during aging, seriously compromise cellular integrity (Stadtman, 1992).

ROS can react with DNA either at the sugar-phosphate backbone or at a base. The former reaction leads to strand breaks, and the latter results in a chemically modified base. The number of hits per day is almost unbelievable. Ames et al. (1993) reported that the number of oxidative hits to DNA per cell per day is about 100,000 in rat and 10,000 in humans. Realizing that DNA repair enzymes are susceptible to oxidation, it is unrealistic to suppose that DNA repair systems would be able to remove all the lesions. Mitochondrial DNA mutates much faster than nuclear DNA, probably because mitochondrial DNA is not covered by histones and is located near the mitochondrial respiratory chain (Richter and Schweizer, 1997). Levels of oxidative damage in mitochondrial DNA of rat liver and human brain cells are at least 10-fold higher than those of nuclear DNA (Shigenaga et al., 1994). The result is an accumulation of damaged DNA during life and impairment of cellular performance owing to dysfunction of cellular processes.

The increasing ROS production during life affects cellular signal transduction systems and, subsequently, the activity of transcription factors and gene expression. It is suggested that alterations in mitogen-activated protein kinase (MAPK) and other signal transduction pathways can contribute to the age-related functional alterations brought about by deregulation of the cellular redox state during aging (Guyton et al., 1997). Subsequently, the expression of a number of genes has been found to be altered in senescent cultured cells and *in vivo* aging. Among these genes are those encoding heat-shock proteins (HSPs), metallothioneins (MTs), and DNA-repair enzymes. In many cases, these changes in gene expression are dependent on a corresponding decline or rise in the activity of transcription factors. The best-characterized transcription factors include activator protein-1 (AP-1) and heat-shock factor (HSF), both of which show decreased activity with aging (Guyton et al., 1997). (For a description of these transcription factors and the above-mentioned MAPK system see Chapters 3 and 2, respectively.) Alterations in the MAP kinase/phosphatase balance in aged rat hepatocytes are correlated with the observed decline in AP-1 DNA-binding activity in these cells (Guyton et al., 1997). A reduced ability to express HSPs in senescent animals is well documented, although the actual cause for the decline in HSP expression has not been determined (Guyton et al., 1997). However, activation of HSF is mediated by thioredoxin (TRX). At too-low TRX concentrations, the redox state is shifted toward an oxidizing microenvironment, and subsequent DNA-binding activity of HSF is reduced (Jacquier-Sarlin and Polla, 1996). This observation could explain the reduced HSF activity in aged animals. On the contrary, the activity of nuclear factor-κB (NF-κB), which controls immune, inflammatory, and acute-phase responses, increases with age. It is believed that alterations in redox state with age also influence NF-κB activity (Guyton et al., 1997). It may therefore be concluded that the redox state is an important factor in changing the activity of genes involved in aging.

8.2.3.2   Involvement of genetic components in aging

During the past 10 years, researchers have been using single gene mutations to unravel the underlying mechanisms affecting aging. In the beginning of this research effort, the nematode *Caenorhabditis elegans* fulfilled a central role in the investigation of the relationship between caloric restriction and aging (Gems and Partridge, 2001). Kenyon et al. (1993) reported on a mutation of the "dauer larva formation" (daf-2) gene, which expresses an important protein of the insulin signaling pathway. A dauer larva is a developmentally arrested larval form that is induced by crowding and starvation and is very long lived. Reduced expression of the mutated daf-2 gene results in an extended life span, identical to the way caloric restriction brings about this effect (Kimura et al., 1997; Pletcher et al., 2002). However, this mutant also showed an increased resistance to oxidative stress by overexpression of cytosolic SOD and CAT (Gems, 1999). Since then, the interaction between life span, caloric restriction, and oxidative stress resistance has been a subject of discussion (Gems and Patridge, 2001; Holzenberger et al., 2003).

Identical mutations were investigated in *Drosophila* species, resulting in, for instance, the "insulin/insulin-like growth factor" (Inr) mutant (Aigaki et al., 2002; Tatar et al., 2001) and the *"methuselah"* (mth) mutant (Lin et al., 1998). The *D. melanogaster* gene Inr is homologous to the mammalian insulin receptor–like genes (Inr-1r) and the *C. elegans* daf-2 gene (Holzenberger et al., 2003; Tatar et al., 2001). A reduced expression of Inr shows a decreased activity of the juvenile hormone and an extended life span (Tatar et al., 2001). Most long-lived Inr and daf-2 mutants develop dwarfism and hypofertility; however, some Inr mutants display normal fertility and growth (Aigaki et al., 2002; Holzenberger et al., 2003). This seems to depend on the level of decreased activity of the mutant gene (Holzenberger et al., 2003).

Lin et al. (1998) isolated the *Drosophila* mutant *methuselah* with a hypomorphic allele for expressing a G protein–coupled receptor (GPCR). This mutant line (mth) displays an approximately 35% increase in average life span and enhanced resistance to various forms of stress. Realizing that GPCRs are involved in signal transduction of hormones and that the redox state changes during life toward an oxidizing environment, which seems to be under hormonal control, it may be speculated that the shift in redox state is controlled via the GPCRs. Thus, a lower gene expression of GPCRs in mth mutants may result in a lower level of signals regulating the redox state in mth than in wild-type *D. melanogaster*.

These examples illustrate the involvement of genetic components in aging and show that many genes are involved in controlling the dynamic and complex aging process (Martin et al., 1996; Pletcher et al., 2002). Although our understanding of the underlying mechanisms controlling aging in *Drosophila* is still limited, it is expected that considerable progress will be made in the near future (Gems and Partridge, 2001), certainly in light of the present knowledge of the *Drosophila* genome.

8.2.3.3   *Environmental stressors contributing to aging*

The previous chapters have extensively described how environmental stressors exert their damaging effects at the cellular level and how the cell reacts by mobilizing stress defense systems, which respond to the stress in a cooperative way. One conclusion is that whatever the type of stress, the effects are observed throughout the cell and whatever the severity of the stress, damage to molecules, structures, and organelles is the result. Therefore, environmental stress contributes to the aging process, and even if the damage were completely removed, the cost of an increased metabolic rate to abolish the stress and repair the damage leads to increased aging. The contribution of environmental stress depends, of course, on the severity and duration of the stress.

Under normal conditions, the cellular stress defense systems contribute to a cellular homeostasis and limit cellular degeneration by a number of activities, of which the most important ones are highlighted below.

The stress-protein system consists of various families of stress proteins (SPs). Members of the sp70 family fulfill a stabilizing role and prevent denaturation of functional proteins or promote refolding. Small heat-shock proteins (SHSPs) are regulators of the redox state and support antioxidant systems (Arrigo, 1998). Ubiquitin, a special stress protein, is involved in cytosolic proteolysis by tagging those proteins that have to be degraded. All these functions are directed toward preventing accumulation of aberrant proteins. Under stressful conditions, priority is given by the cell to these activities by allowing a fast induction of extra stress proteins, at the expense of the induction of normal proteins. In this way, the accumulation of damaged proteins is minimized.

Production of ROS is a phenomenon that also occurs under steady state conditions. Consequently, antioxidant scavenging and detoxification systems are active under these conditions. They scavenge and detoxify ROS to a certain extent, because low concentrations of ROS (e.g., of $O_2^{\bullet-}$ and H_2O_2) are necessary as messengers or even for protection, for instance, termination of lipid peroxidation by $O_2^{\bullet-}$. Thus, under steady state conditions a balanced operation by antioxidant systems, such as SOD and CAT, is necessary. This also applies to the level of vitamin C. At too-high concentrations, vitamin C may act as a pro-oxidant and reduce ferric ions, which, in reduced form, may catalyze lipid peroxidation reactions (Yu, 1994). The same applies to too-high concentrations of iron. Iron that is not used as a catalyst is therefore stored in ferritin. However, the body's iron content increases with age, and it is suggested that this accumulation increases the risk of oxidative damage (Beckmann and Ames, 1997). As a consequence of an ongoing lipid peroxidation, the lipolytic action of phospholipase A_2 (PLA$_2$) is stimulated. It was demonstrated that there is a correlation between an age-related increase in PLA$_2$ content of microsomal membranes and an increase in oxidatively altered lipid hydroperoxides (Yu, 1994). It was also shown that PLA_2 activity in the inner mitochondrial membrane increases in response to conditions associated with increased oxidant production (Shigenaga et al., 1994).

A gradual shift in redox state with age toward an oxidizing environment is an underlying factor in the aging process. It has been observed that the GSH content, which determines to a great extent the cellular redox state, decreases with age, whereas the lipid peroxide concentration increases with age (Yu, 1994). Whether this inverse correlation between the lower content of GSH and the higher level of lipid peroxides represents a causal relationship has not yet been established (Yu, 1994).

Arking (1998) selected *D. melanogaster* strains over 22 generations for extended longevity, and ascertained a significantly higher activity of the antioxidant defense systems and a corresponding delay in senescence from day five of adult life. The enzymes tested were SOD, CAT, and xanthine dehydrogenase (XDH). Arking (1998) also ascertained that there seems to be a close correlation between enhanced resistance to oxidative stress, lower levels of oxidative damage, and increased longevity, as predicted by the "oxidative stress theory of aging." This theory demonstrates the important role of antioxidants in the degeneration process of cells.

In a recent article, Arking (2001) reconfirmed these test results but made some additional observations that are of interest regarding aging. Sister strains of *D. melanogaster* with identical longevity phenotypes do not require molecular equivalence. One strain primarily expressed genes for SOD, whereas the sister strain mainly depended on genes overexpressing CAT. The organism may have multiple genetic strategies to cope with similar levels of oxidative stress. It was further noticed that the extended longevity did not result in a reduced metabolic rate and that these animals did not comply with the "rate of living hypothesis" (Arking, 2001). Although this points to a seemingly improved fitness, it seems to negatively affect other stress defense systems, and it seems that the wild-type *Drosophila* has found an optimum level reflecting the highest fitness under average conditions. The antioxidant defense system plays a very important role in aging and extended longevity.

The mixed function oxygenase (MFO) system, containing cytochrome P450 enzymes, constitutes another cellular stress defense system. It converts both endogenous and exogenous compounds of a hydrophobic character into intermediate compounds of a hydrophilic character for further degradation. However, in executing this task, it fulfills a dual role: During this biotransformation process, reactive intermediates may be generated, at which point the superoxide anion radical is formed and cellular macromolecules are damaged. Such superoxide radical formation is triggered by redox cyclers, such as quinonoid compounds and paraquat. In this way, the MFO system contributes unintentionally to the aging process by consuming part of the antioxidant capacity.

The metal stress defense system is formed by MT. The principal roles of MT lie in the regulation of the metabolism of essential trace metals, such as copper and zinc, and the detoxification of other heavy metals, such as cadmium, by means of sequestration. There is increasing evidence, however, that there is another role for MT, that is, as a free radical scavenger (Bauman et al., 1991; Sato and Bremner, 1993; Zhang et al., 2001). In this way, this defense system contributes to the regulation of aging under normal conditions. Under stressful conditions, heavy metals, such as cadmium, may cause a variety of effects and responses, ranging from induction of MT, GSH, proto-oncogenes, and stress proteins to inhibition of zinc-containing enzymes, generation of oxidative stress and lipid peroxidation, and inhibition of DNA repair enzymes. Thus, heavy metal concentrations in excess of MT detoxification capacity may damage all kinds of macromolecules and may contribute considerably to cellular degeneration.

Under normal conditions, the basal signal transduction systems probably participate in the genetic control of senescence by conveying extracellular signals to the transcriptional machinery in the nucleus (or according to the negative reasoning, by limiting signal transduction). Under stressful conditions, a stress response is triggered by the same signals. In this way, the basal signal transduction systems also play an important role in minimizing cellular damage and contribute to a slower aging process.

Environmental stress contributes to the aging process by increased cellular deterioration. The stress defense systems mentioned above try to abolish these effects, but even if they do, it is at a cost. The induction of the stress response and the degradation of exogenous compounds requires energy, generated by increased metabolism of energy carriers. This not only causes extra production of ROS and subsequent negative effects on cellular performance, but in the long term it may shorten the life span, according to the rate of living hypothesis.

8.3　Cell death by environmental stress

8.3.1　Introduction

If cells are injured by environmental stress, there are several options for responding. Depending on the severity of the stress, the cell will try to repair the damage and, if successful, it will return to homeostasis. It may be possible that the damage can only be partly repaired, but that the cell will stay alive under suboptimal conditions. Another option is that, in the case of DNA damage, and depending on the cell cycle, the cell will enter a process leading to growth arrest, implying G_1 or G_2 phase. This process may be reversible pending repair (Gill and Dive, 2000) or irreversible and, although the cell stays alive, it is genetically dead (Evan and Littlewood, 1998). If the severity is too high, the cell dies, either by necrosis, a passive process, or by apoptosis, an active process also called "programmed cell death" (PCD). The various options are illustrated in Figure 8.1.

Figure 8.1 provides an overview of various pathways and options the cell uses to respond to endogenous signals and to cellular stress. The

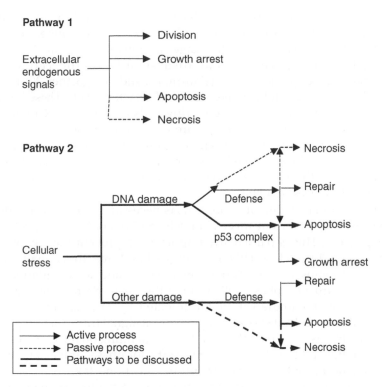

Figure 8.1 Overview of various pathways and options the cell uses to respond to endogenous signals (pathway 1) or to cellular stress (pathway 2). Cellular stress pathways are divided into pathways activated by DNA damage and by other damage because of a different cellular response. The pathways indicated by thick lines are discussed in this section. (Partly adapted from Evan and Littlewood, 1998.)

pathways of cellular stress are subdivided into two groups: those that are activated by DNA damage and those that are activated by other damage. This differentiation is made because of different cellular responses to these damages. Because the aspects of defense and repair have been described in the previous chapters, emphasis will be put on events leading to necrosis or apoptosis included in pathway 2 of Figure 8.1. The pathways to be discussed in this section are indicated in the figure by thick lines. The pathways activated by endogenous signals and the growth arrest pathway within the cellular stress group of pathways are considered outside the scope of this book and will only be referred to if necessary.

Necrosis is induced by severe environmental disturbances and is characterized by swelling of the cytoplasm and cytoplasmic organelles, early rupture of the plasma membrane, clumping of the chromatin, and usually swelling of the nucleus (Lee and Shacter, 1999). Necrosis is a passive and therefore rather slow process that leads to spillage of intracellular contents. This spillage may result in inflammatory reactions by neighboring tissues. Apoptosis is regarded as an active and progressive response to physiological and pathological stimuli (Lee and Shacter, 1999), as well as to damage by environmental stress. It is characterized by early and prominent condensation of nuclear chromatin, cell shrinkage, fragmentation of nuclear DNA in internucleosomal particles, and disassembly of the cell in membrane-enclosed apoptotic vesicles (Kaufmann and Hengartner, 2001; Krauss, 2001), also called apoptotic bodies (Lee and Shacter, 1999). These bodies are removed by phagocytosis by neighboring cells of the tissue (Gill and Dive, 2000). These morphological changes are observed during the final stage of degradation. The early reversible changes after a lethal injury include cytoplasmic blebbing. This change is common to all types of cell death progression. However, the blebs observed in apoptosis often contain organelles (Trump and Berezesky, 1995).

Under normal conditions, apoptosis is necessary to achieve an adequate balance between sufficient survival of cells and overwhelming proliferation and expansion. This is particularly important to prevent malignant growth (Fuchs et al., 1997). Jabs (1999) distinguishes three functionally distinct phases in apoptosis: an induction, an effector, and a degradation phase. During the induction phase, cells receive the death-promoting signal, such as growth-factor withdrawal, or are affected by environmental stressors, such as oxidants or heat shock. Many different signal transduction pathways participate in this induction phase of PCD, depending on the type of stimulus or stressor. Beyond this stage, during the effector phase, universal regulatory events including anti- and pro-apoptotic functions take place and bring together the initial diverse signaling processes into a few stereotyped pathways. Here, the cells pass a point of no return and are irreversibly condemned to death (Jabs, 1999). So far, the paths of necrosis and apoptosis are identical, but somewhere downstream from this point of no return, they diverge, leading to a different degradation process and morphological changes. We speculate that the cell tries to keep control of the degradation process, resulting in an apoptotic cell death, but if, for reasons to be explained in the following section, the cell is not able to finish it properly, it ends up in an uncontrolled process of cellular deterioration: necrosis.

The processes of apoptosis and necrosis are complex, and many factors and systems are involved, influencing the process in a pro- or anti-apoptotic direction. These processes will be described in the following section. Emphasis will be put on the induction of apoptosis by environmental stressors. Since signals from these stressors are mixed up with endogenous signals, they will be discussed together. Analyzing these cell-death processes, attention will be paid to the role of the cellular stress defense systems in cell death. If the cell is not able to defend itself properly, then it initiates PCD and eventually dies. The process of apoptosis is cell specific, however, and the pathways and mechanisms described below are not uniformly applicable to all cell types (Gill and Dive, 2000).

8.3.2   Mechanisms and processes initiating and effectuating cell death

As explained in the Introduction, cell death is caused by endogenous and exogenous factors. These factors initiate cell death in a different way. Endogenous factors, such as the tumor necrosis factor (TNF)-α, convey their signal via a receptor-mediated pathway, also called death-receptor-triggered apoptosis (Krauss, 2001). Exogenous factors, such as environmental stressors, initiate cell death by causing cytotoxic stress and cell injury. Thus, the initiation of cell death by endogenous and exogenous factors is different; they follow different pathways in the induction phase, but they finally converge in the effector phase. Therefore this phase is discussed first.

The effector phase is characterized by the induction of proteases, phospholipases, and endonucleases, which all participate in the degradation of cellular constituents, leading to the breakdown of the cell either in an apoptotic or a necrotic way (the degradation phase).

With regard to the proteolytic network, central components of the effector path are a certain class of proteases, the caspases. The caspases mediate degradation of a number of key structural and housekeeping proteins and execute cell death (Krauss, 2001). The caspase family consists of initiator and effector caspases that are present in an inactive form. The initiator caspases receive pro-apoptotic signals and initiate the activation of a caspase cascade. The initiator caspases are activated by interaction with a trans-membrane receptor or by cytotoxic influences (Krauss, 2001). The trans-membrane receptors are activated by endogenous signals, such as TNF-α, and convey signals controlling development, survival, and death (Wertz and Hanley, 1996). The signals from cytotoxic stress will be discussed below in detail.

Caspases are not the only group of enzymes that execute apoptotic cell death, since nucleases, phospholipases, and protein kinases may also participate, but, according to Salvesen and Dixit (1997), caspases are absolutely required for the accurate and limited proteolytic events that characterize apoptosis. More recent publications also refer to caspase-independent pathways leading to apoptosis (Zimmermann et al., 2002). It is questionable, however, whether one can still speak of apoptosis or whether a kind of intermediate process between apoptosis and necrosis is taking place. Caspases also participate in necrotic cell death, but they are not essential for this process (Jabs, 1999; Leist et al., 1997). Caspase inhibitors may block some or all of the apoptotic morphology, but in that case the cell will ultimately die by a slower nonapoptotic cell death (Green and Reed, 1998).

The induction pathway activated by cellular stress is very complicated and implicates nuclear and mitochondrial functioning, signal transduction systems, and second messengers. First, the role of mitochondria in the apoptotic process, in relation to oxidative stress by reactive oxygen species, is discussed. Next, the "p-53-pathway" in relation to DNA damage (see Figure 8.1) is examined. This section concludes with an explanation of the roles of Ca^{2+} and HSPs in cell death.

8.3.2.1   The role of mitochondria in the stress-mediated pathway
Upon exposure to environmental stressors, ROS can accumulate to toxic levels. At these elevated concentrations, ROS lead to oxidative stress resulting in cytotoxicity and damage to cell structures and DNA and improper functioning of organelles, such as mitochondria (Storz and Polla, 1996). These effects may initiate a series of events that culminate in apoptosis (Flores and McCord, 1997; Fuchs et al., 1997) or necrosis (Richter and Schweizer, 1997).

Green and Reed (1998) describe three kinds of mechanisms in mitochondria that contribute to the decay of mitochondria and to the induction of apoptosis or necrosis:

1. Disruption of electron transport, oxidative phosphorylation, and ATP production
2. Alteration of cellular redox state
3. Release of proteins that trigger the activation of caspases

8.3.2.1.1   Disruption of electron transport, oxidative phosphorylation, and ATP production.   Disruption of electron transport may be brought about by inhibitors of enzymes of the electron transport chain (ETC). These inhibitors may be exogenous and intracellular compounds, such as ceramide, a second messenger involved in apoptosis signaling (Hannun, 1996; Hannun and Luberto, 2000). This messenger inhibits cytochrome *c* in the ETC (Green and Reed, 1998). The consequence of suboptimal functioning of the ETC is increased ROS production, which may affect the permeability of the inner membrane of mitochondria. This results in loss of the proton gradient over the inner membrane and subsequent impairment of oxidative phosphorylation and loss of ATP production (Chaudère, 1994). ATP appears to be required for downstream events in apoptosis (Green and Reed, 1998). Leist et al. (1997) showed that ATP generation, either by glycolysis or by mitochondria, was required for the active execution of the final phase of apoptosis (see Section 8.3.3). Aside from a loss of ATP production, improper functioning of the ETC results in increased $O_2^{\bullet-}$ and subsequent H_2O_2 production. Quillet-Mary et al. (1997) demonstrated that these ROS act as an early major indicator of apoptosis. The pathways followed by these ROS to exert their effects are discussed below.

8.3.2.1.2   Alteration of cellular redox state. $O_2^{\bullet-}$ and H_2O_2 can be scavenged by GSH in mitochondria, yielding glutathione disulfide (GSSG) species and a subsequent shift of the redox state toward an oxidizing environment. Under severe oxidative stress conditions, this may lead to the release of mitochondrial Ca^{2+}, disturbing Ca^{2+} homeostasis (Jabs, 1999). Ca^{2+} re-uptake decreases the proton gradient over the inner membrane and subsequent ATP production. Thus, disturbance of mitochondrial Ca^{2+} homeostasis could be a starting point in the process leading to apoptosis (Fuchs et al., 1997; Goossens et al., 1995; Richter and Schweizer, 1997).

Moreover, oxidative stress in mitochondria can promote extra-mitochondrial activation of the transcription factors nuclear factor-κB (NF-κB) and activator protein -1 (AP-1). The activation of these transcription factors is believed to be mediated by a shift in the cellular redox state owing to mitochondrial ROS production and subsequent GSH depletion (Quillet-Mary et al. 1997). This topic is discussed in more detail below.

8.3.2.1.3   Release of proteins that trigger the activation of caspases. The disorder in mitochondrial functioning is the cause of the opening of a large conductance channel, known as the mitochondrial permeability transition (PT) pore. Its constituents include both inner membrane proteins, such as the adenine nucleotide translocator, and outer membrane proteins, such as porin, a voltage-dependent anion channel. They operate in concert, presumably at inner- and outer-membrane contact sites, and create a channel through which molecules up to 1.5 kDa may pass. PT pore opening leads to swelling of the matrix owing to hyper-osmolarity and possible rupture of the outer membrane of mitochondria (Green and Reed, 1998).

The opening state of the PT pore may be controlled by intra- and extra-mitochondrial signals. Mitochondrial ROS production and GSH depletion may trigger the opening of the PT pore from the inside. However, signals from the outside, such as a sustained increase in cytosolic Ca^{2+} or activation of caspases via the receptor-mediated pathway, may also promote PT pore opening. Thus, ROS and thereby alterations of the cellular redox state are important signaling components of the induction phase (Jabs, 1999).

Tanguay et al. (1999) reported on the role of SHSPs as negative regulators of cell death through modulation of the cellular redox state. It is speculated that SHSPs modulate redox control by indirectly enhancing intracellular GSH content. It is believed that unphosphorylated SHSPs, forming large aggregates, embody GSH (Arrigo, 1998; Mehlen et al., 1996a,b, 1997). Enhanced GSH levels afforded by SHSPs could explain the contribution of SHSPs to delaying or preventing cell death.

Mitochondria undergoing PT pore opening release at least two apoptogenic proteins, one of which is a caspase-activating protein, cytochrome *c*. The other factor is the apoptosis-inducing factor (AIF). Once released into the cytosol, cytochrome *c* induces the activation of pro-caspases, in cooperation with other cytosolic cofactors, such as apoptosis protease activating factor 1 (Apaf 1) (Desagher and Martinou, 2000; Jabs, 1999; Krauss, 2001). The *Drosophila* homolog is called Apaf 1–related killer, ARK (Zimmermann et al., 2002) or Dark (Rodriguez et al., 2002). The ATP-dependent activation occurs in a complex of caspases, Apaf 1 protein, and cytochrome *c*, which is also known as the apoptosome (Krauss, 2001). AIF processes the pro-caspases (Green and Reed, 1998) but also causes the condensation of chromatin and DNA fragmentation (Jabs, 1999). Although cytochrome *c* release is frequently observed in vertebrates, it appears not to be a universal aspect of apoptosis. In contrast, there is currently no evidence that cytochrome *c* takes part in apoptosis in either nematodes or insects (Green and Reed, 1998; Zimmermann et al., 2002).

Promotion of PT pore opening from the outside, for example, by activated caspases, which in turn can induce caspase activation, can be considered a feedforward amplification loop (Green and Reed, 1998). As a kind of all-or-nothing switch, the PT pore can be considered the central coordinator and executioner of cell death. PT pore opening seems to be the point of no return (Jabs, 1999).

Because of the consequences of PT pore opening and activation of caspases, this process is extra protected and controlled by a group of proteins called the B-cell lymphoma/leukemia-2 (Bcl-2) family of proteins. Some, such as Bcl-2 itself, inhibit cell death, while others, such as Bax, promote cell death. Many of these proteins interact with each other through a complex network of homo- and heterodimers, with one monomer antagonizing or enhancing the function of another. In this way, the ratio of pro- and anti-apoptotic Bcl-2 family proteins may determine the likelihood of the cell undergoing cell death (Jabs, 1999). Localized in the outer mitochondrial membrane, Bcl-2 and its homologs have been suggested to directly interact with the PT pore complex, thereby modulating its opening. Bcl-2 serves as a caspase substrate and is thereby converted to a Bax-like factor (Jabs, 1999). The converstion of Bcl-2 to Bax by caspase demonstrates the irreversible effect of caspase activation, irrevocably leading to cell death.

Another antagonist of the apoptotic process appears to be Raf-1. This enzyme is a member of the MAPK cascade upstream in the MAPK kinase (MEK)– extracellular signal-regulated kinase (ERK) (classical) pathway (see Figure 2.2). However, Raf-1 also interacts with the pro-apoptotic, stress-activated "apoptosis signal-regulating kinase 1" (ASK 1). This interaction allows Raf-1 to act independently of the MEK-ERK pathway to inhibit apoptosis. ASK 1 is an important mediator of apoptotic signaling initiated by a variety of death stimuli, including TNF-α, Fas activation, oxidative stress, and DNA damage (Chen et al., 2001).

8.3.2.2   DNA damage and the pathway leading to cell death

If DNA is damaged by genotoxic stress, a partly different pathway leading to apoptosis is followed. In this pathway, the tumor suppressor protein p53 plays a central role. Activation of this protein induces either a stable growth arrest or apoptosis (Polyak et al., 1997; Ryan et al., 2001). Here, only the apoptotic pathway will be discussed in relation to DNA damage.

If cells are exposed to alkylating agents, ionizing radiation, or free radicals, DNA may be damaged, for example, in the form of strand breaks. Poly (ADP-ribose) polymerase (PARP) will then bind quickly to the lesion to prevent accidental recombination and to signal to DNA-damage response pathways (Tong et al., 2000). PARP is an abundant nuclear protein, present in most eukaryotic cells. Upon binding, PARP undergoes automodification, leading to the formation of long, branched poly (ADP-ribose) polymers. It uses nicotinamide adenine dinucleotide $(NAD)^+$ as a substrate, which may cause the depletion of intracellular NAD^+ and which may have serious consequences for the energy metabolism in the cell. Subsequent dissociation of the PARP enzyme facilitates the DNA repair machinery to repair the lesion (Tong et al., 2000).

The activity of the PARP enzyme also leads to increased p53 levels and activation of this protein by phosphorylation (Gottifredi et al., 2000; Zhao et al., 2000). p53 mediates cell cycle arrest or apoptosis through its ability to transcriptionally activate a selected number of genes (Lohrum and Vousden, 2000; Zhao et al., 2000). Polyak et al. (1997) reported on a group of 13 p53-induced genes, some of which are involved in redox control. Zhao et al. (2000) refer to expression of some genes in relation to apoptosis. One of these genes, Apo-1/Fas, is believed to produce the trans-membrane receptor involved in cell cycle and cell death signaling. Krauss (2001) reported that variants of the p53 protein are involved in the redistribution of the Fas death receptor from the cytosol to the cell membrane. This is considered an alternative pathway for inducing apoptosis without transcriptional activation (Krauss, 2001).

Returning to the DNA-damage induced p53 pathway leading to apoptosis, Polyak et al. (1997) postulate that this activity involves a three-step process: (1) the transcriptional induction of redox-related genes, (2) the formation of reactive oxygen species, and (3) the oxidative degradation of mitochondrial components, culminating in cell death. This three-step process describes an integration of the p53 activity with the stress-mediated pathway described in the previous section. The basis for this integration is the creation of a shift in redox state by expression of redox-related genes, including NADPH:quinone oxidoreductase, which is a potent generator of ROS (Polyak et al., 1997). Moreover, Zhao et al. (2000) reported that, as a result of p53 transcriptional activity, the activity of superoxide dismutase (SOD) is turned off. This will drastically increase $O_2^{\bullet-}$ levels. Apart from this intra-mitochondrial activity emerging from the transcriptional activity of the p53 protein, Krauss (2001) reports that this activity is enhanced by activation of the Bax protein, one of the pro-apoptotic proteins of the Bcl-2 family of proteins controlling mitochondrial pore opening. It is even speculated that Bax not only has a controlling function, but that an increased Bax concentration leads to the formation of pores in mitochondria (Krauss, 2001). This would mean that p53 also regulates the induction of stress-related apoptotic processes via cytosolic activities. It also suggests that the process induced by p53 is under strict enzymatic control, which increases the probability that the death process results in apoptosis.

This is in contrast to degeneration processes initiated by environmental stressors, such as inhibitors of electron transport in mitochondria or redox cyclers. These processes depend primarily on the concentration of the stressor and the capacity of antioxidant systems. In the latter case, the antioxidant systems defend the cell against degeneration to the extent possible. However, in the case of activation of ROS production by the p53 system, the cell has already chosen the apoptotic pathway, and an active antioxidant system does not fit in this program. Therefore, the role of the antioxidant system in the apoptotic pathway, initiated by p53, is minor. This is exemplified by the shut off of SOD. It also demonstrates that this p53 pathway differs from the stress-mediated pathway with regard to control, notwithstanding the fact that the biochemical mechanisms applied inside mitochondria are the same. The actions in the stress-mediated pathway and in the p53-pathway described above are summarized in Figure 8.2.

8.3.2.3   The role of Ca^{2+} in cell death from environmental stress

Intracellular Ca^{2+} is present in different concentrations in the cytosol, the nucleus, the endoplasmic reticulum (ER), and in mitochondria. In these compartments and organelles, it fulfills different functions. Ca^{2+} contributes to cellular integrity, and disturbance of Ca^{2+} homeostasis affects cellular structures and processes. This section will focus on the effects of these disturbances on the processes leading to cell death.

As explained in the section on the stress-mediated pathway, mitochondria play an important role in the induction phase of apoptosis or necrosis, by the production of ROS and the subsequent decline of cellular ATP level and change of the redox state. In these processes, two aspects of the involvement of mitochondrial Ca^{2+} homeostasis are important: Ca^{2+} release into the cytosol and Ca^{2+} cycling.

The production of $O_2^{\bullet-}$ and H_2O_2 is partly scavenged by GSH, which is regenerated by NADPH, yielding NAD^+ as end-product of the regeneration step. Normally, NAD^+ is regenerated to NADH in the mitochondrium in the Krebs cycle, but at excessive ROS production, it may also be hydrolyzed to ADP-ribose, which stimulates, in the presence of several cofactors, the release of Ca^{2+} from mitochondria. ADP-ribose is a substrate for ADP-ribosylation of inner membrane proteins. This reaction has a dual effect: It influences membrane permeability and subsequently the proton gradient over the mitochondrial inner membrane, and ADP is no longer available for regeneration. This promotes NADH depletion (Gardner et al., 1997; Richter and Schweizer, 1997). The release of Ca^{2+} probably takes place through a Ca^{2+}/H^+ antiporter, so a PT pore opening is not required for it (Richter and Schweizer, 1997). It also means that Ca^{2+} release from mitochondria is an early event during the induction phase. Second, re-uptake of Ca^{2+} follows a different pathway, which leads to a cycling process of Ca^{2+} over the inner membrane, affecting the proton gradient over the inner membrane and subsequently decreasing ATP production (Richter and Schweizer, 1997). The final result is a deregulation of all major mitochondrial functions with regard to energy generation and redox control, which culminates in PT pore opening and subsequent effects.

The adverse effects of ROS on mitochondria may increase cytosolic Ca^{2+} concentration (Gardner et al., 1997). Both ROS and cytosolic Ca^{2+}, supported by a shift in cellular redox state toward an oxidizing environment, may activate signal transduction pathways and protein kinase systems, which may lead to the expression of apoptotic death genes through nuclear factors, such as NF-κB and Jun/Fos (Gardner et al., 1997). Cytosolic Ca^{2+} concentration can be further increased by activation of phospholipase C (PLC), which

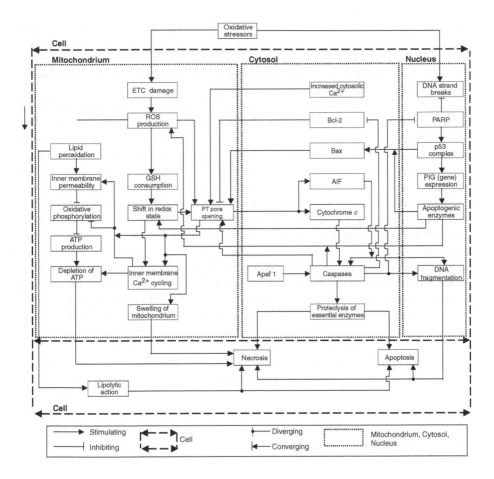

Figure 8.2 Integration of cell-death pathways induced by environmental stressors in mitochondria and nuclear DNA. In mitochondria, impairment of the functioning of the electron transport chain causes production of reactive oxygen species with downstream effects on adenosine triphosphate production, glutathione consumption, and mitochondrial Ca^{2+} homeostasis, resulting in a redox shift and membrane pore opening. Nuclear DNA damage activates a p53 protein complex leading to transcriptional activation of redox-related enzymes, which induce a production of reactive oxygen species in mitochondria. At this point the two pathways converge. Pore opening in the mitochondrial membrane leads to activation of caspases by release of caspase-activating proteins. Caspases are the effectors of cellular degradation, which results in either apoptosis or necrosis. Abbreviations: ATP = adenosine triphosphate; ETC = electron transport chain; ROS = reactive oxygen species; GSH = glutathione; PT = permeability transition; Bcl-2/Bax are members of the Bcl-2 family of proteins controlling mitochondrial pore opening; AIF = apoptosis-inducing factor; Apaf1 is a cofactor in caspase activation; PARP = poly (ADP-ribose) polymerase; PIG = p53-induced genes; p53 = a protein complex activating apoptogenic genes.

leads to the release of Ca^{2+} from the ER. Moreover, a re-uptake into the ER via Ca^{2+}-ATPase pumps is impaired by reduced ATP levels. The same applies for Ca^{2+} in nuclei. Nuclei also take up Ca^{2+} in an ATP-dependent way, because there is no free diffusion of ions through nuclear pores (Trump and Berezesky, 1995).

Increase of cytosolic Ca^{2+} precedes prelethal changes (Yu et al., 2001) and involves reduced energy metabolism. Although a correlation between an increased cytosolic Ca^{2+} level and some physical symptoms, such as cytosolic shrinkage and blebbing of the cytoplasm and nucleus during the early stage of apoptosis, has not been established yet, Trump and Berezesky (1995) speculate that Ca^{2+} induces at least the phenomenon of shrinkage. They believe that increased cytosolic Ca^{2+} levels activate Ca^{2+}-dependent Cl^- channels in the plasma membrane, with subsequent loss of this ion and water. Furthermore, the effect of cytosolic Ca^{2+} on bleb formation is generally recognized.

In the effector phase, after mitochondrial pore opening, the mitochondrial matrix may swell from the penetration of sucrose (Richter and Schweizer, 1997) accompanied by water owing to hyper-osmolarity of the matrix (Green and Reed, 1998). The contribution of cytosolic Ca^{2+} is a calcification step within the already swollen mitochondria, resulting in the formation of calcium hydroxyapatite. The swollen mitochondrium is one of the phenomena of a characteristic state of the cell entering the final phase of degradation (Trump and Berezesky, 1995). In this degradation phase, cytosolic and nuclear Ca^{2+} participate in the induction of phospholipase A_2 (PLA_2), protein kinases, and endonucleases. Together with the activation of special proteases by activated caspases, the activation of all these enzymes results in the breakdown of cellular structures and cellular fragmentation, either in an apoptotic or a necrotic way.

8.3.2.4   The role of HSPs in cell death

It is known that HSPs protect cells from damage by environmental stress, but what is their role in preventing the induction of the apoptotic process or stopping it? For instance, in previous sections it has been described how SHSPs contribute to the maintenance of the cellular redox state and delay or prevent the induction of the apoptotic process indirectly. Mehlen et al. (1996b) reported on a more direct role for hsp27, a member of the SHSP family; that is, under certain conditions, hsp27 protects cells from Fas-mediated apoptosis, a process stimulated by endogenous signals.

The most abundantly inducible HSP of the sp70 family, hsp70 is also able to prevent or block apoptosis. Mosser et al. (1997) observed that heat-induced apoptosis was blocked in HSP-expressing cells. Heat stress activates sphingomyelin in membranes, resulting in the production of ceramide. This second messenger activates the proto-oncogene c-Jun via the stress-activated protein kinase (SAPK)/c-Jun N-terminal kinase (JNK) pathway of the MAPK cascade. We presume that this cascade is activated by ROS as a consequence of cytochrome *c* inhibition by ceramide. Activated c-Jun, in its turn, initiates the apoptotic process by indirectly activating caspases, especially caspase-3, which cleaves PARP enzymes. This effect induces the p53-death-pathway and subsequently apoptosis (Mosser et al., 1997). Mosser et al. (1997) noticed that hsp70 can interrupt this chain of processes at two points: First, high levels of hsp70 are able to inhibit signaling events upstream from SAPK/JNK; second, hsp70 is able to inhibit the activation of caspase-3 and block the apoptotic process (Beere and Green, 2001; Mosser et al., 1997). This cascade of effects has been summarized in Figure 8.3.

8.3.3 Apoptosis or necrosis?

The description of apoptosis and necrosis, as presented in the previous sections, indicates that, at least during the induction phase, there are common pathways for both processes. It further appears that, if the injury cannot be repaired, the cell aims at an apoptotic cell death, because this exerts the least damage to surrounding tissues. When matters get out of control, the death process diverges from the apoptotic pathway and ends up in a degenerative process leading to a necrotic cell death. The question then remains of where the pathway diverges and what the reasons are for this split.

To answer the latter question, Gardner et al. (1997) investigated this problem by exposing specific cell types to different H_2O_2 concentrations. They concluded that the type of cellular response to H_2O_2 exposure depends on the concentration of the oxidant. Above a certain concentration, the cell dies through necrosis, while moderate levels of H_2O_2 induce apoptosis. Lee and Shacter (1999) are of the opinion that the type of cell death is also dependent on cell type and that, under certain conditions, H_2O_2 can even inhibit apoptosis: In the presence of relatively low concentrations of H_2O_2, the calcium ionophore, A23187, was unable to induce apoptosis in Burkitt's lymphoma cells, whereas this was possible in the absence of H_2O_2. They found that, under specific conditions, H_2O_2 inhibits apoptosis by depleting the cell of ATP. They suggest that, at low H_2O_2 concentrations, the drop in ATP is transient, allowing the cell to die through apoptosis. At higher H_2O_2 concentrations, the drop in ATP appears to be irreversible, thus leading to a necrotic cell death. Leist et al. (1997) observed the same effects, however, using different cell types and oxidants. They found that, under these specific conditions, a residual ATP level above 50% was sufficient to change the mode of cell death from necrosis to apoptosis and that the ATP generated by glycolysis was sufficient to execute the apoptotic process.

ATP is required in the apoptotic death process for the execution of processes catalyzed by caspases. Experiments reported by Leist et al. (1997) demonstrate that ATP is also required for the process of nuclear collapse, the ordered packaging of chromatin, and DNA degradation before cell lysis, which is affected by activated caspases. Proper execution of this process seems to be essential for continuing and finishing the apoptotic process (Zhang and Xu, 2002).

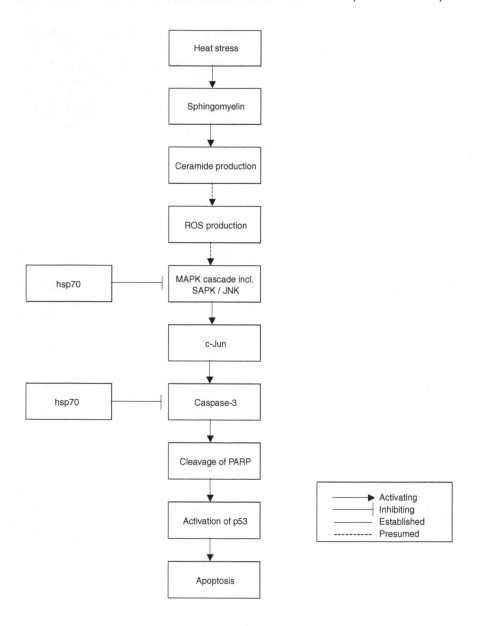

Figure 8.3 The role of hsp70 in heat-induced apoptosis. Heat stress induces a cascade of actions leading to apoptosis; hsp70 is able to interrupt or prevent this apoptotic process at two points: first, upstream from the SAPK/JNK enzymes of the MAPK system; second, by inhibiting the activation of caspase-3, a protease of the caspase family affecting apoptosis. Abbreviations: ROS = reactive oxygen species; MAPK = mitogen-activated protein kinase; SAPK = stress-activated protein kinase; JNK = c-Jun N-terminal kinase; c-Jun = proto-oncogene product; PARP = poly (ADP-ribose) polymerase; p53 = a protein complex activating apoptogenic genes; hsp = heat-shock protein. (Adapted from Mosser, D.D. et al., *Mol. Cell. Biol.* 17: 5317–5327, 1997.)

These results imply an answer to the question of where the pathway leading to necrosis diverges from the apoptotic pathway: An improper, uncontrolled degradation of DNA owing to too-low ATP concentrations seems to be the cause of the cell entering the necrotic pathway. This means that the switch from an apoptotic pathway to a necrotic pathway takes place during the degradation phase and that the two processes may have a common pathway during the induction and effector phase.

This conclusion applies for a moderate stress level. At very high severity, when many more things get out of control at the same time, the switch to a necrotic cell death may take place much earlier, for instance, by the rupture of mitochondrial or plasma membranes. Another reason could be that, under severe conditions, the drop in ATP is so substantial that ATP-dependent activation of caspases is impaired. In that case a switch to a necrotic process may already have taken place during the effector phase. Depending on incomplete execution of the apoptotic death process, there may be various kinds of morphological differences in cell death. In this range of physical differences, apoptosis and necrosis can be considered two extremes of a continuum of cell death phenomena (Jabs, 1999; Leist et al., 1997; Yu et al., 2001).

chapter 9

The integrated cellular stress defense system

9.1 Introduction

In the previous chapters, the major stress defense systems have been described as more or less independent systems, each responsible for its specific tasks. The purpose of this chapter is to demonstrate that these systems form one integrated cellular defense system against the different insults by various environmental stressors. Analyzing this supposition, three important questions arise:

1. What is specific in the individual systems that makes each system different from the others?
2. What are the most important stress effects leading to a common reaction of all or most of the defense systems?
3. How and to what extent do these systems cooperate so that they indeed can be considered one integrated system?

9.1.1 Specific characteristics of the individual systems

The following facts can be observed when analyzing the systems discussed in the previous chapters:

1. The basal signal transduction systems are not specific defense systems; they function as common tools in normal cellular processes and, under stressful conditions, as intermediates between stressor and stress response.
2. Proteotoxic stress and the stress-protein systems are characterized by stressors causing denaturation of proteins and the defense system responding by induction of stress proteins.
3. Oxidative stress is caused by oxygen free radicals generated by components of the oxidative stress response system during the

processing of oxidants. Oxidative stress is mainly characterized by the effects of lipid peroxidation.

4. Metal stress is specific by its type of stressor, mainly heavy metals, and the metal response system is characterized by its detoxifier and scavenger metallothionein (MT).
5. The speciality of the mixed-function oxygenase (MFO) system, in the context of environmental stress, is the degradation of xenobiotics.

It may therefore be concluded that the systems differ from each other in these aspects.

9.1.2 Important stress effects

The most important common stress effects brought about by various stressors are:

1. Disturbance of structural integrity by damage to proteins, enzymes, or lipids. The net results are loss of structural integrity and increased permeability of membranes, leading to impaired functioning of ion pumps and channels and subsequently to disturbance of sodium and calcium homeostasis.
2. Inhibition of enzymes by denaturation, oxidation, or metal substitution, resulting in improper functioning of enzyme systems.
3. The effects mentioned in points 1 and 2 together result in the inhibition of respiration and of adenosine triphosphate (ATP) synthesis by oxidative phosphorylation. A switch to anaerobic energy production is ultimately necessary to restore the physiological $NAD^+/NADH$ (nicotinamide adenine dinucleotide hydride) ratio and to prevent interruption of glycolysis.
4. Oxidation of thiol groups leads to depletion of reductants, such as glutathione (GSH) and thioredoxin (TRX), and shifts the cellular redox state toward an oxidizing environment. This may lead to protein-S-thiolation and increased NADPH consumption. To compensate for these effects, the cell will activate its pentosephosphate shunt and related metabolic processes, at the expense of glycolysis and Krebs cycle operation.
5. These effects also influence the working of basal signaling systems which may, depending on the severity of the stress, result in either activation or inhibition of cell proliferation.

Taken together, all these effects are central in disturbing cellular homeostasis, and each of them may become so severe that the cell enters the process leading to apoptosis; however, these effects also trigger activation of the defense systems, which may lead to common responses.

9.1.3   Cooperation and integration of systems

Defense systems respond to and assist one another in the following ways:

9.1.3.1   Relationship between the stress-protein system and the oxidative stress response system

The stress-protein system reacts to disturbances of structural integrity and damage of proteins and enzymes, irrespective of the type of stressor. What matters is aggregation of aberrant proteins. For instance, in the case of protein damage by oxidative stress, for example, caused by the tumor necrosis factor-α (TNF-α), a prominent generator of superoxide anion radical, hydrogen peroxide, or menadione, expression of small heat-shock proteins (SHSPs) leads to a decrease in reactive oxygen species (ROS) and an increase in GSH level. Thus, the protective activity of SHSPs against oxidative stress may result from their ability to increase the intracellular content of GSH, which in turn, decreases ROS levels. The mechanism by which SHSPs are able to increase GSH content is unknown, but it is speculated that unphosphorylated SHSPs forming large aggregates interact with GSH, promoting GSH storage (Mehlen et al., 1996a, 1997). It is believed that, in this case, storage could be looked upon as a form of sequestration of part of the cellular GSH content. This part would not be directly involved in redox control owing to its embodied condition in SHSP aggregates.

Another possibility is that, because SHSPs behave as protein chaperones, they act as specialized chaperones toward enzymes, such as glutathione reductase, involved in the ROS-downgrading pathways (Arrigo, 1998). Either way, enhanced GSH levels afforded by SHSPs could explain the contribution of SHSPs, not only to oxidative stress response but also to delaying apoptosis as a subsequent effect. Although the major cytosolic stress protein, hsp70, is also induced by hydrogen peroxide (Burdon et al., 1990), it seems to not play a role in ROS reduction but rather inhibits phospholipase A_2, thereby stabilizing membrane structures (Mehlen et al., 1996a). Thus, this description provides two examples in which stress proteins assist the oxidative stress defense system to respond to oxidative stress.

9.1.3.2   Relationship between the oxidative stress response system and the metallothionein system

The oxidative stress response system is not only supported by components of the stress-protein system but also by MT of the metal response system. MT assists in scavenging free radicals, such as the hydroxyl radical, and thus reduces lipid peroxidation (Bauman et al., 1991; Sato and Bremner, 1993; Zhang et al., 2001). The exact mechanism by which MT protects against oxidative stress is unknown, but scavenging appears to involve the metal-thiolate clusters of MT, and any damage to MT could be repaired by reduced GSH (Bauman et al., 1991). In this process, zinc would be released which, in turn, has a stabilizing effect on membranes and may limit lipid peroxidation. This might be brought about by free zinc reducing iron uptake by the cell

(Sato and Bremner, 1993). Although the concentration of heavy metals is a major determinant of cellular MT levels, induction of MT synthesis also occurs by chemical stimuli producing oxygen free radicals. Notwithstanding the fact that Sato and Bremner (1993) observed a correlation between hepatic MT concentrations and lipid peroxide levels, they also found that lipid peroxidation is not an essential event for MT synthesis during oxidative stress.

Depletion of glutathione also did not affect the synthesis of MT (Cherian, 1995). Experiments reported by Dalton et al. (1996) showed that in mouse, the transcription factor MTF-1 (metal-dependent transcription factor 1) plays a key role in regulating MT gene expression in response to oxidative stress. Mechanisms of activation of MTF-1 are not well understood, however. Dalton et al. (1996) also confirmed in their experiments that the transcription factor is reversibly activated in response to free zinc levels. Zinc released by scavenging of oxygen free radicals therefore plays a dual role in protection against oxidative stress: membrane stabilization and MT induction. See the hypothetical model of Roesijadi (1996) shown in Figure 5.2. Depending on the concentrations of and affinities among the compounds in this model, it could describe the induction of MT by oxidative stress, in which zinc plays a central role.

9.1.3.3   Relationship between the metallothionein system and other stress response systems

The metal response system cooperates with the stress-protein system. Goering et al. (1993) demonstrated that after acute exposure of hepatic tissue to cadmium, *de novo* synthesis of stress proteins was one of the earliest changes occurring. Acute cadmium toxicity may cause conformational changes in proteins by interacting with sulfhydryl groups and may inhibit enzymes. Accumulation of aberrant proteins could therefore be a mechanism activating heat-shock factor (HSF) and inducing stress protein synthesis. Specific induction by cadmium of glucose-regulated proteins (GRPs), members of the stress protein family, may involve the property of this heavy metal to disrupt cellular calcium homeostasis (Goering and Fisher, 1995). Thus, the induced stress proteins are involved in protecting the cell against proteotoxicity rather than in rendering the stressor harmless. Note that the role of MT in scavenging oxygen free radicals, as described in the previous section, seems to be different: MT assists the oxidative stress response system, but in the case of exposure to cadmium, which inhibits detoxifying enzymes such as superoxide dismutase (SOD), leading to increased concentrations of ROS (Beyersmann and Hechtenberg, 1997; Goering et al., 1995), the role of MT in scavenging these radicals is partly a matter of removing the damage caused by cadmium. Similarly, elevated cadmium levels stimulate induction of GSH (Sugiyama, 1994) to remove the effects brought about by cadmium in their covalent bonds with thiol groups. This could be considered a form of assistance of the oxidative stress response system to the metal stress response system. Beyersmann and Hechtenberg (1997) observed that in cadmium-treated cells, the induction of synthesis of protective sulfhydryl compounds precedes lipid peroxidation and DNA damage. Thus, it is an early response.

9.1.3.4   Relationship between the mixed-function oxygenase system and other stress response systems

There is only an indirect relationship between the MFO system and other stress response systems. Such an indirect relationship is shown in Figure 6.2. Both xenobiotics and oxidants can activate the same genes, mainly related to the MFO system. The MFO system has a role in assisting cellular defense by transforming xenobiotic compounds into hydrophilic intermediates that can be excreted after conjugation. In this way, these xenobiotics are made harmless, and possible toxic effects, such as formation of DNA and protein adducts, are prevented. At the same time, however, intermediate products of this biotransformation process, such as semiquinones, may be harmful because of their involvement in redox cycling and consume the scavenging capacity of other stress defense systems. An active MFO system may therefore have both positive and negative effects on cellular homeostasis and other defense systems.

9.1.3.5   Summary

The above-mentioned facts give an impression of the most important relationships and ways of cooperation between the defense systems in their efforts to protect the cell against possible damage by environmental stressors. The effects discussed and the responses of the individual defense systems analyzed in previous chapters, and the interrelationships between the individual systems as discussed above are summarized in Figure 9.1. Even this complicated scheme confines itself to major effects on structures, processes, and factors affecting cellular homeostasis and to related responses by the defense systems. It demonstrates an intricate network of effects, in which a particular type of stressor brings about damage to all major parts of the cell, comprising the cellular integrity and functioning. Consequently, all the defense systems become involved in their response to the effects caused by the particular stressors. Hence protection of cellular integrity requires the defense systems to cooperate. Basal signal transduction pathways have a coordinating task with regard to signaling effects and evoking responses and therefore play an essential role in accomplishing this cooperation. This is made possible not only because the different response systems use their own specific transcription paths, but also because promoters of their genes are equipped with a variety of responsive elements, such as heat-shock elements (HSEs), metal-responsive elements (MREs), antioxidant responsive elements (AREs), and xenobiotic responsive elements (XREs). This also enables the stress response system to respond to stress signals in a cooperative but differentiated way.

9.2   The central hypothesis

In Chapter 1 of this book, the hypothesis was postulated that the individual defense systems form an integrated cellular defense system. From all the information presented and analyzed, it may be concluded that the defense systems at least cooperate to maintain cellular integrity. The main reason for this is that a particular stressor may bring about effects initiating responses of all the stress defense systems discussed. For example, an oxidant causing oxidative stress by generating ROS subsequently causes damage not only to membrane lipids but also to proteins and enzymes throughout the cell, and not only induces antioxidants to respond to oxidative stress but also initiates responses of the stress-protein system and the metal response system. Thus, oxidative stressors also directly affect other stress response systems. From this point of view, the cooperation between the defense systems is a consequential phenomenon and does not reflect a strategic plan.

The question remains of whether these defense systems also act as an integrated system. In cellular systems, integration does not refer to a high-level decision center but rather to coordination in the cooperation of the stress defense systems. When analyzing the complex scheme in Figure 9.1, it is striking that among the many factors influencing and controlling cellular homeostasis, only a few are pivotal in both exhibiting stress effects and initiating stress response. The following four aspects seem to be the major factors that absorb the signals from stress effects and transfer them into signals for stress response:

1. Damage to proteins and enzymes
2. Disturbance of redox state
3. Changes in free zinc levels
4. Effects on signal transduction systems

9.2.1   Damage to proteins and enzymes

Almost all environmental stressors are able to damage proteins and enzymes throughout the cell, including structural proteins, enzymes of metabolic processes, DNA synthesis and repair, as well as proteins and enzymes of defense systems including antioxidant systems, MT, and enzymes of the MFO system. The accumulation of these damaged proteins triggers the induction of stress proteins to prevent further damage, to help restore what can be repaired, and to remove what has to be degraded. This new situation can only be temporary because new proteins and enzymes have to be synthesized to restore balances. Thus, the stress-protein defense system functions as a central system to respond to damage throughout the cell. The system provides a general response within the total defense by the cell, whereby the signal is integrated and channeled through accumulation of damaged proteins.

9.2.2   Disturbance of redox state

The redox state is mainly determined by the balance of reduced and oxidized thiols. Oxidative stressors oxidize thiols, which affect the cellular redox state if consumed GSH is not replenished fast enough. The TRX system described in Chapter 4 also restores modified thiols to their unmodified reduced state at the expense of NADPH. This also applies to the regeneration of oxidized GSH. To provide sufficient reducing power in the form of NADPH, activation of the pentosephosphate shunt may be necessary at the expense of ATP synthesis. A switch to a more oxidizing environment makes other thiol-containing proteins and enzymes more susceptible to further insults. This may lead to induction of other stress response systems, including the stress-protein system and the metal-response system. Here, one can observe a true example of integration of effects and a coordinated response because all signals are regulated by the cellular redox, which to a great extent is determined by the GSH/GSSG ratio.

9.2.3   Changes in free zinc levels

Free zinc probably controls MT induction to some extent which, in turn, not only protects against metal stress but also scavenges radicals, thus reducing damage to proteins, lipids, and nucleic acids. Generation of free zinc is effected not only by metal stress but also by oxidants, the effect of which is influenced by the redox state of the cell. This is another example of the way stress signals from different sources are integrated, resulting in a coordinated response by free zinc as pivotal factor.

9.2.4   Effects on signal transduction systems

The overall response is mainly effected by synthesis of extra compounds participating in the defense systems, including stress proteins, antioxidants, P450 enzymes, and MT. To some extent, this is initiated by activation of transcription factors and related gene expression via the basal signal transduction systems. They are mainly activated by oxidant, metal, and genotoxic stress. The systems may amplify the signal to the transcription factors but may also influence each other negatively, like the effect of cAMP on the MAPK system. Altogether, participation of basal signal transduction systems may increase the stress response and, at the same time, inhibit cellular processes like proliferation. Thus, stress signals from various sources are also integrated and transduced via these systems, leading to combined responses by the stress defense systems. This analysis also supports including the basal signal transduction systems in the discussion of the typical stress defense systems.

9.2.5   *Summary*

Taking these analyses together, it can be concluded that to some extent, one can speak of an integrated response by the different cellular stress defense systems, providing support for the hypothesis that they work together as an integrated cellular stress defense system.

The major factor through which stress effects are integrated and translated into a coordinated stress response is probably the redox state, substantiated to a great extent by the GSH/GSSG ratio. This factor also influences other important factors, such as free zinc level and basal signal transduction, which are made more susceptible to activation by a shift of the redox state to an oxidizing environment. The redox state can thus be considered a pivot in integrating the response of the individual stress defense systems.

The intricate network of cellular stress defense mechanisms contributes to homeostasis, not only at the cellular level, but also at the level of the whole organism. Since stress is such an important phenomenon in nature, all organisms have evolved stress defense mechanisms, and these systems therefore form a key factor in the relationship between an organism and its environment. Future studies using simultaneous analysis of whole-genome responses will undoubtedly shed more light on the patterns explored in this book.

chapter 10

Summary

Animals may be subjected to environmental stress caused by physical stressors, such as heat, or chemical substances in the environment. These stressors may affect the functioning of the animal and evoke responses on both the organismal and cellular level. At the cellular level, it may cause a disturbance of homeostasis which, depending on the severity of the damage, may be restored by cellular responses; partly restored, leading to a suboptimal functioning of the cell; or cannot be restored, leading to cell death. The effects are differentiated in primary effects at the molecular level, secondary effects on cellular structures and processes, and tertiary effects, such as a collapse of structures or a switch to a different metabolic pathway. Depending on the type of stressor, different forms of cellular stress may be developed. In this book, the stresses were divided into proteotoxic, oxidative, metal, and genotoxic stress.

Proteotoxic stress, for example, caused by heat, leads to denaturation of proteins resulting in damage of cellular structures, varying from effects on membrane permeability to a collapse of filament structures. Denaturation of proteins may further lead to an inhibition of enzyme activity resulting in a disturbance of respiration and metabolism. Accumulation of denatured proteins evokes the induction of stress proteins, which protect essential proteins and enzymes against further damage by means of stabilization. Stress proteins also participate in repairing damage by promoting protein refolding and contribute to the degradation of aberrant proteins by means of their chaperoning role. For all these functions there are several groups of stress proteins, which may be constitutively present or may be induced by stress. For the induction, multiple pathway systems exist, which are activated according to the type of stress and the localization of the damage. The hearts of these pathways are formed by special transcription factors binding specific sequences on the promoter of the genes expressing stress proteins. An example of such a specific pathway for induction of specific stress proteins is ionophores. These stressors may disturb Ca^{2+} homeostasis in the endoplasmic reticulum. This may upset the glycosylation process in this organelle and induce the expression of glucose-regulated proteins (GRPs), a special group of stress proteins.

Some chemicals, such as redox cyclers, may cause oxidative stress by generating reactive oxygen species (ROS), such as the superoxide anion radical. This radical may cause damage by oxidation of essential sulfhydryl groups in enzymes and antioxidants. This may result in a disturbance of respiration, energy generation, and other metabolic processes, as well as a shift in cellular redox state toward an oxidizing environment. At the processing of superoxide anion radicals, harmful compounds, such as the hydroxyl radical, are generated. These compounds may cause lipid peroxidation, leading to membrane damage and cytotoxic effects by reaction products. The hydroxyl radical further causes DNA damage, which may be crucial for cellular survival. To prevent all these harmful effects, the cell uses both enzymatic antioxidant systems, such as superoxide dismutase (SOD), catalase (CAT), and glutathione peroxidase (GSH-P$_x$), and antioxidant scavengers, such as glutathione (GSH), TRX, and vitamins. Oxidative stress may evoke the induction of antioxidants via basal signal transduction systems, such as the cyclic adenosine monophosphate route, the Ca^{2+} second messenger, and the mitogen-activated protein kinase (MAPK) cascade. Combined signals (positive and negative) may result in the activation of specific transcription factors and stress response genes.

Heavy metals are the most important factors causing the so-called metal stress. The severity of the stress depends on the type of metal and its cellular concentration. This book refers to metal stress by cadmium and copper when cells are exposed to high concentrations of these metals. Cadmium may oxidize sulfhydryl groups in proteins and enzymes and may substitute other metals in catalytic domains of enzymes. This may lead to damage of structural proteins and inhibition of enzymatic functions, including those of SOD, CAT, and DNA repair enzymes. Inhibition of these essential enzyme systems leads to oxidative and genotoxic stress, resulting in cytotoxicity. Copper also induces oxidative and genotoxic stress, but because it is a transition metal, the pathways followed and the mechanisms applied may be different. For instance, Cu$^+$ may be directly involved in lipid peroxidation by producing hydroxyl radicals. However, Cu$^+$ is not involved in inhibiting CAT activity. DNA damage may also be caused in a more direct way by oxidation of both thymine residues and backbone molecules. Cadmium induces backbone cleavage in a more indirect way and inhibits DNA repair enzyme activity.

Protective measures against metal stress are induced at various metal concentrations. For instance, at low cadmium concentrations, MT and GSH are induced. At higher, noncytotoxic concentrations, stress proteins are induced as well, probably by proteotoxicity. This can be considered an emergency measure. Extra MT is induced under these conditions to detoxify the metal by binding to the metal and rendering it harmless by sequestration. The MT is induced by activation of metal responsive transcription factors, for example, via free zinc removed from MT and enzymes. MT can also be induced, however, by activation of other transcription factors via the basal signal transduction systems. Probably, it is this combination of activated pathways and different transcription factors that leads to a metal-specific response in terrestrial arthropods by the induction of specific MT. For instance, snails probably show both metal- and tissue-specificity by inducing a cadmium isoform of MT in the midgut gland and a copper isoform in the mantle. *Drosophila* have tissue specificity for metals, and their isoforms of MT show preference for different localizations within a cell.

Special stressors may develop genotoxic stress by causing DNA strand breaks. Examples are arsenite, hydrogen peroxide (by generating hydroxyl radicals), and ultraviolet light (probably by generating singlet oxygen species). Many organic chemicals present in the environment may cause genotoxic stress by forming DNA adducts. This may happen either directly, for instance, by quinonoid compounds, or indirectly, for instance, by metabolites of polycyclic aromatic hydrocarbons (PAHs), formed upon biotransformation of these compounds by P450 enzymes of the MFO system. Quinonoid compounds may also be activated by the same system in a one-electron reduction process, at which point semiquinones are formed. These may then generate superoxide anion radicals by autoxidation, giving rise to oxidative stress. The different stressors induce specific P450 iso-enzymes of the MFO system.

These iso-enzymes also vary among species and phyla. For instance, of the six major classes of P450 enzymes in insects, only one class includes sequences from vertebrates. Foreign compounds inducing P450 enzymes are divided in two classes because they induce different iso-enzymes via two different pathways. They are classified as phenobarbital (PB)–type inducers or 3-methylcholanthrene (3MC)–type inducers. Moreover, the two pathways for the induction of P450 enzymes by foreign compounds seem to be interrelated with basal signal transduction systems. These foreign compounds are also able to activate non-P450 genes in this way. These correlations also explain to some extent the relationship between the MFO system and the oxidative stress response system, that is, how oxidized intermediates of the mixed function oxygenase system initiate the induction of oxidative stress response enzymes.

All these stress defense systems also fulfill important functions during normal cellular life. The actions undertaken and the mechanisms applied are basically identical under both stressful and normal conditions. The response during stress requires intensified actions at the expense of normal processes. This also applies to the basal signal transduction systems. Although seemingly not belonging to the stress defense systems, they play an important role in transducing stress signals to the transcriptional machinery and at the same time suppressing processes such as cell proliferation. Detailed analysis in this book shows that most of the different stressors exert not only special effects but also important side effects causing various types of stress. This leads to responses by multiple cellular stress defense systems. Stress effects and the initiation of stress responses appear to be mainly channeled via a limited number of regulating systems, predominantly influenced by the following four aspects:

1. Damage to proteins and enzymes
2. Disturbance of redox state
3. Changes in free zinc levels
4. Effects on signal transduction systems

Among these aspects influencing regulating systems, the GSH/GSH disulfide ratio, mainly determining the redox state, plays a pivotal role in maintaining cellular homeostasis. Because the effects from different stresses culminate in the aspects listed above, they become integrated and stress defense responses become coordinated. Hence, the individual systems cooperate in their response. It may therefore be concluded that there is substantial support for the hypothesis that the individual stress defense systems mentioned above form an integrated cellular stress defense system with interrelated connections, leading to a cooperative response. It does not mean that they always amplify each other's activity. It may even lead to a down-regulation of part of the integrated system, which is not required. In this way, costs are saved. This intricate network of effects and responses is summarized in Figure 10.1.

If cells are not able to withstand environmental stress, they may enter a pathway leading to apoptosis, an active process, also called "programmed cell death." Three phases in the process leading to cell death are distinguished: an induction, an effector, and a degradation phase. Induction of apoptosis by environmental stressors may follow two pathways. First, environmental stressors may cause mitochondrial disturbances, mainly by disturbing the respiration process and related generation of adenosine triphosphate (ATP). This may lead to depletion of this energy carrier and GSH, the main contributor to the maintenance of the cellular redox state. A disturbance of Ca^{2+} homeostasis, together with the other factors, leads to mitochondrial pore opening, through which molecules are released that initiate the effectuation of the apoptotic process. The alternative pathway is activated if DNA is damaged by genotoxic stress. Strand breaks initiate the activation of a p53 complex, which leads to expression of apoptogenic genes, mainly influencing the cellular redox state by affecting mitochondrial processes in a way comparable with what was described above. The two pathways converge in the mitochondrium.

The effector phase is characterized by the activation of so-called caspases by the activator molecules released from mitochondria. Activation of these effector molecules can be considered as a point of no return. A cascade of reactions takes place, including disturbance of cytosolic Ca^{2+} homeostasis and the activation of proteases, nucleases, and lipolytic enzymes. The degradation phase is characterized by condensation of chromatin, cell shrinkage, fragmentation of DNA, and disassembly of the cell in apoptotic bodies. These are removed by phagocytosis. If the cell is not able to finish this process properly, it ends up in an uncontrolled deterioration process: necrosis. The process of necrosis is passive and characterized by swelling of the cytoplasm, nucleus and organelles and an early rupture of the plasma membrane. This

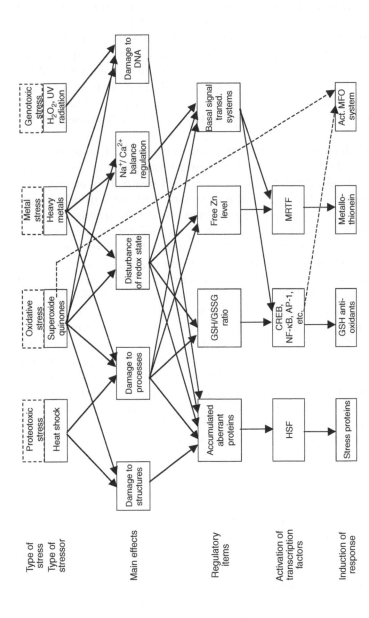

Figure 10.1 Summary of the intricate network of effects and responses caused by environmental stress, showing the various stressors exert their effects throughout the cell. The signals are integrated by some key regulation factors and transformed in a coordinated response. Solid lines indicate established pathways, dashed lines represent presumed but less well-established events. Abbreviations: GSH = glutathione; GSSG = glutathione disulfide; H_2O_2 = hydrogen peroxide; HSF = heat-shock factor; CREB = cAMP-responsive element-binding protein; cAMP = cyclic adenosine monophosphate; NF-κB = nuclear factor-κB; AP-1 = activator protein-1; MRTF = metal-responsive transcription factor.

may result in inflammatory reactions by neighboring tissues. The controlled process of apoptosis requires energy in the form of ATP, while an improper, uncontrolled degradation of DNA, owing to depletion of energy, seems to cause the cell to enter the necrotic pathway. At very high severity of the stress, when many more things get out of control at the same time, the switch to a necrotic cell death may take place before DNA degradation.

When the damage is only partly repaired and the stress persists over long periods of the animal's life, a kind of suboptimal dynamic equilibrium is formed by developing tolerance to the stress. This may have local effects in some tissues of the organism, but it may also influence the functioning of the whole organism, culminating in effects on life-history traits of the animal, such as survival, fecundity, and life span. Life span is determined by the aging process.

Although the effects of aging are exhibited at the individual level in the form of senescence, the basic process of aging takes place at the cellular level as reflected in cellular deterioration and cell death. It is an ongoing process during the whole life span of the organism, with slow-dividing cells seeming to have a greater impact on aging than fast-dividing cells. Changes in cellular processes and functions during aging probably have a genetic basis, resulting in changes in the activity of transcription factors and gene expression. A central role in the initiation of these changes is attributed to a shift in the redox state toward an oxidizing environment, making the cell more susceptible to oxidation by an increased production of ROS. The oxidation involves lipids, proteins, and nucleic acids and leads to the accumulation of damaged molecules during the cell's life. The accumulation is progressive and is the manifestation of cellular degeneration, contributing to the aging process. Environmental stressors contribute to the aging process through their negative effects on cellular performance and the cost of defense. Many theories contribute to the explanation and analysis of the aging process. Most important in relation to aging by environmental stress is the "free radical theory of aging," which holds that free radical reactions are the underlying factors responsible for the aging process. It is therefore not surprising that the cellular stress defense systems play an important role in retarding the aging process by (partly) eliminating the effects of environmental stress.

Sublethal stress may often lead to the development of tolerance to the stress. If the tolerance has a genetic basis, it is inheritable. Tolerance may also be acquired, however, by physiological adjustments to the stress. This type of tolerance is lost after exposure to the stress. It cannot be passed on to offspring. This acquired tolerance is called acclimatory tolerance. If, among individuals, the genetic variation underlying the physiological adjustments is robust enough, selection on this property may lead to populations having inheritable tolerance to the related stress.

These different types of tolerance have been discussed in this book in relation to the activation of the cellular stress response system. In this context, heat tolerance plays a major role. If cells are exposed to a heat shock, stress proteins are induced to prevent or repair possible damage by the stress. At the same time, resistance to further heat shocks is built up. This acquired tolerance results in reduced recovery times after exposure and enhanced survival, and it correlates with the accumulation of different types of stress proteins. After exposure to a heat shock, these heat-tolerant cells further show a reduced inhibition of protein translation, a diminished disruption of the cytoskeleton, and a faster redistribution of heat-shock proteins than non-heat-tolerant cells. It is believed that faster kinetics of the heat-shock proteins underlie these effects. Moreover, HSPs protect themselves against denaturation by their interaction with other proteins. The costs of the acquired heat tolerance are reflected in reduced fecundity.

In some aspects, tolerance to cold stress shows identical properties. To survive winter conditions, insects produce small carbohydrates, such as glycerol and sugars, and antifreeze proteins to lower their melting point and supercooling point. Two strategies may be followed: tolerating or avoiding freezing. Freeze-tolerant insects accept freezing of extracellular fluid and let this happen at mild freezing temperatures to minimize risk of membrane damage. Freeze-intolerant species avoid freezing by decreasing their supercooling point to the extent necessary. At cold shock or chilling, insects build up tolerance by cold hardening at a temperature well above their freezing point. During this process, a fast induction of antifreezes occurs. The mechanisms involved in the development of cold tolerance by *Drosophila* are not well known. It seems that the induction of stress proteins is not an issue in these cold-hardening processes.

Resistance to insecticides may have a genetic basis or may be inducible, sometimes stimulated by plant toxins in an insect–host plant relationship. In all cases, cytochrome P450 enzymes of the MFO system are involved. This book discussed examples of both inducible and inheritable tolerance to organophosphorous and pyrethroid insecticides. The case study on *Drosophila* species in the Sonoran Desert demonstrates how substrate-specific resistance to cactus species was developed by the induction of specific cytochrome P450 enzymes.

Finally, tolerance to heavy metals in relation to the MT system has been reviewed. In *Drosophila*, tolerance to heavy metals is tissue-specific and to some extent metal-specific, for example, copper is mainly absorbed in cuprophilic cells of the midgut, zinc and copper in the Malpighian tubules, and cadmium in midgut epithelium cells. Acclimatory tolerance is achieved by sequestration or excretion. Sequestration includes both the induction of MT and compartmentalization of heavy metals. Inheritable tolerance to heavy metals in *Drosophila* involves adaptation by either gene amplification or selection on a variation of regulatory genes for the expression of MT. Adaptation shows increased fitness effects as a result of increased fecundity, survivorship, and adult weight in polluted environments, but at the cost of reduced growth and reproduction in unpolluted environments. It confirms evolutionary theories that adaptation increases fitness, but only in relation to the circumstances to which the animal has adapted itself.

In conclusion, this book demonstrates how an intricate network of cellular stress-defense systems contributes to diminishing the effects of environmental stress, not only at the cellular level but also at the level of the whole organism. The systems are constitutively present, but their activity is increased under stressful conditions by the induction of special components of the defense systems, depending on the type and severity of the stress. This induction is under strict control, however, not only to prevent energy costs, but also to prevent detrimental effects resulting from inducing stress defense compounds under normal conditions. It appears to be a balanced operation, which may increase fitness under special circumstances or may prevent irreversible damage by stress to cells and, ultimately, to the whole organism. These defense systems therefore form a pivotal factor in the relationship between an organism and its environment.

Abbreviations

AC	adenylate cyclase
ADP	adenosine diphosphate
AFP	antifreeze protein
AHH	aryl hydrocarbon hydroxylase
AhR	aryl hydrocarbon receptor
AhRE	Ah-regulatory element
AIF	apoptosis-inducing factor
AMP	adenosine monophosphate
AP-1	activator protein-1
Apaf 1	apoptosis protease activator factor 1
ARE	antioxidant responsive element
ARK	A paf 1-related killer
Arnt	Ah-receptor nuclear translocator
ASK 1	apoptosis signal-regulating kinase
ATP	adenosine triphosphate
BaP	benzo[*a*]pyrene
BaP-O	oxidized benzo[*a*]pyrene
Bcl-2	B-cell lymphoma/leukemia-2 protein
BiP	binding protein
bp	base pair
calm	calmodulin
CaMK	Ca^{2+}-calmodulin-dependent protein kinase
cAMP	cyclic adenosine monophosphate
CAT	catalase
CdK4	cyclin-dependent protein kinase 4
CDR-1	cadmium-responsive protein-1
c-fos/jun/myc	cellular proto-oncogenes
CRE	cAMP-responsive element
CREB	cAMP-responsive element-binding protein
CYP	cytochrome P450
cys	cysteine residue
daf	dauer larva formation gene

DAG	diacylglycerol
DIA	diaphorase
DJNK	c-Jun N-terminal kinase in *Drosophila* cells
dMTF-1	metal-responsive transcription factor in *Drosophila* cells
DNA	deoxyribonucleic acid
DTT	dithiothreitol
eIF-2α	eukaryotic initiation factor-2α
Em1p	a transmembrane protein in the ER
EpRE	electrophile response element
ER	endoplasmic reticulum
ERAS	errors, radicals, antioxidants, and scavengers
ERK (p42-p44)	extracellular signal-regulated kinase
ETC	electron transport chain
Fos/Jun	proto-oncogene products forming AP-1
GAGA	basal promoter binding element/protein
GAPDH	glyceraldehyde-3-P-dehydrogenase
GPCR	G protein-coupled receptor
G6PDH	glucose-6-phosphate dehydrogenase
GR	glucocorticoid receptor
GRE	glucocorticoid response element
grp(GRP)	glucose regulated protein
GSH	glutathione
GSH-Px(m)	glutathione peroxidase (membrane-bound isoform)
GSSG	glutathione disulfide
GTP	guanosine triphosphate
HOG 1	high osmolarity gene 1 product
HS	heat shock
Hsc(HSC)	heat-shock cognate
HSE	heat-shock element
HSF	heat-shock (transcription) factor
hsp(HSP)	heat-shock protein
Igf-1r	insulin-like growth factor-1 receptor gene
I-κB	inhibitor-κB
IM	inner membrane
INA	ice nucleating agent
Inr	insulin/insulin-like growth factor gene
IP$_3$	inositol triphosphate
JH	juvenile hormone
JNK (p46-p54)	c-Jun N-terminal kinase
kDA	kilo-Dalton
LEP	life energy potential
LMW	low-molecular-weight
LPR	learn pyrethroid resistant
LT	lymphotoxin
LT50s	time taken for half the sample to be killed
MAO	monoamine oxidase

MAPK	mitogen-activated protein kinase
MAPKAP-K1/2	MAPK-activated protein-kinase 1/2
MBF	metal responsive element binding factor
3-MC	3-methylcholanthrene
MDA	malonyldiadehyde
MDP	methylenedioxyphenyl
MDR	multidrug resistance
MEK	MAPK kinase (MAPKK)
MEKK	MEK kinase (MAPKKK)
Mep	metal element binding protein
MFO	mixed function oxygenase
Mito	mitochondrium
MKK3/4	MAPK kinase 3/4
MKP	MAPK phosphatase
MLSP	maximal life-span potential
MRE	metal-responsive element
mRNA	messenger ribonucleic acid
MRTF	metal-responsive transcription factor
MT	metallothionein
MTF-1	metal-responsive transcription factor in human cells
mth	methuselah
mt-hsp70	mitochondrial-hsp70
MTI	metallothionein transcription inhibitor
Mtn/Mto	genes in *Drosophila melanogaster* expressing MT
MTN/MTO	metallothioneins in *Drosophila melanogaster*
MW	molecular weight
NAC	N-acetyl-L-cysteine
NAD (H)	nicotinamide adenine dinucleotide (hydride)
NADP (H)	nicotinamide adenine dinucleotide phosphate (hydride)
NF-κB	nuclear factor-κB
8-OHdG	8-hydroxydeoxyguanosine
OM	outer membrane
PAH	polycyclic aromatic hydrocarbon
p53	a protein complex activating apoptogenic genes
PARP	poly (ADP-ribose) polymerase
PB	phenobarbital
PCB	polychlorinated biphenyl
PCD	programmed cell death
PCDD	polychlorodibenzo-p-dioxin
PDI	protein disulfide isomerase
pGP	p-glycoprotein
PIG	p53-induced genes
PKA	protein kinase A
PKC	protein kinase C
PLA$_2$	phospholipase A$_2$
PLC	phospholipase C

PMA	phorbol 12-myristate 13-acetate
PP	pentosephosphate
PPI	peptidyl-prolyl isomerase
PT	permeability transition
Raf/Ras	enzymes of the MAPK system
RCH	rapid cold-hardening
RER	rough endoplasmic reticulum
RK (p38)	reactivating kinase
RNA	ribonucleic acid
RNP	ribonucleoprotein
RNS	reactive nitrogen species
ROS	reactive oxygen species
rRNA	ribosomal RNA
RTK	receptor tyrosine kinase
RTP	receptor tyrosine phosphatase
SAPK	stress-activated protein kinase
SCAV	scavenger radical
SCAVH	reduced scavenger
SCP	supercooling point
SEK	SAPK kinase
SER	smooth endoplasmic reticulum
shsp/SHSP	small heat-shock protein
snRNP	small nuclear ribonucleoprotein
SOD	superoxide dismutase
sp(SP)	stress protein
α-T·	α-tocopheroxy radical
TATA	basal promoter box
TBP	TATA-box binding protein
TCDD	tetrachlorodibenzo-*p*-dioxin
TF	transcription factor
TFIID	TATA-box binding protein complex
α-TH	α-tocopherol
THP	thermal hysteresis protein
TMQ	2,3,5,6-tetramethylbenzoquinone
TNF-α	tumor necrosis factor-α
TRX	thioredoxin
Ub	ubiquitin
UPR	unfolded protein response
UPRE	unfolded protein response element
UPRF	unfolded protein response (transcription) factor
UV	ultraviolet light
VDAC	voltage-dependent anion channel
XDH	xanthine dehydrogenase
XRE	xenobiotic responsive element

References

Adams, M.D., Celniker, S.E., Holt, R.A., Evans, C.A., Cocayne, J.D. et al. (2000). The genome sequence of *Drosophila melanogaster*. *Science* 287: 2185–2195.

Aigaki, T., Seong, K.-h., and Matsuo, T. (2002). Longevity determination genes in *Drosophila melanogaster*. *Mech. Ageing Dev.* 123: 1531–1541.

Ames, B.N., Shigenaga, M.K., and Hagen, T.M. (1993). Oxidants, antioxidants, and the degenerative diseases of aging. *Proc. Natl. Acad. Sci. USA* 90: 7915–7922.

Ananthan, J., Goldberg, A.L., and Voellmy, R. (1986). Abnormal proteins serve as eukaryotic stress signals and trigger the activation of heat shock genes. *Science* 232: 522–524.

Andrews, G.K., Bittel, D., Dalton, T.P., Hu, N., Chu, W., Daggett, M., Li, Q., and Johnson, J. (1999). New insights into the mechanism of cadmium regulation of mouse metallothionein-1 gene expression. In: *Metallothionein IV* (Klaassen, C.D., Ed.). Birkhäuser, Basel, Switzerland, pp. 227–232.

Arking, R. (1998). Molecular basis for extended longevity in selected *Drosophila* strains. *Curr. Sci.* (Bangalore) 74: 859–864.

Arking, R. (2001). Gene expression and regulation in the extended longevity phenotypes of *Drosophila*. *Healthy Ageing Funct. Long.* 928: 157–167.

Arrigo, A.-P. and Landry, J. (1994). Expression and function of the low-molecular-weight heat shock proteins. In: *The Biology of Heat Shock Proteins and Molecular Chaperones* (Morimoto, R.I., Tissières, A., and Georgopoulos, C., Eds.). Cold Spring Harbor Laboratory Press, Cold Spring Harbor, New York, pp. 335–373.

Arrigo, A.-P. (1998). Small stress proteins: chaperones that act as regulators of intracellular redox state and programmed cell death. *Biol. Chem.* 379: 19–26.

Arthur, J.R., Bremner, I., Morrice, P.C., and Mills, C.F. (1987). Stimulation of peroxidation in rat liver microsomes by (copper, zinc) metallothioneins. *Free Rad. Res. Commun.* 4: 15–20.

Avruch, J., Zhang, X.-F., and Kyriakis, J.M. (1994). Raf meets Ras: completing the framework of a signal transduction pathway. *Trends Biochem. Sci.* 19: 279–283.

Baler, R., Zou, J., and Voellmy, R. (1996). Evidence for a role of hsp70 in the regulation of the heat shock response in mammalian cells. *Cell Stress Chaperones* 1: 33–39.

Barja, G. (2002). Rate of generation of oxidative stress-related damage and animal longevity. *Free Rad. Biol. Med.* 33: 1167–1172.

Baud, V. and Karin, M. (2001). Signal transduction by tumor necrosis factor and its relatives. *Trends Cell Biol.* 11: 372–377.

Bauer, M.F., Hofmann, S., Neupert, W., and Brunner, M. (2000). Protein translocation into mitochondria: the role of TIM complexes. *Trends Cell Biol.* 10: 25–31.

Bauman, J.W., Liu, J., Liu, Y.P., and Klaassen, C.D. (1991). Increase in metallothionein production by chemicals that induce oxidative stress. *Toxicol. Appl. Pharmacol.* 110: 347–354.

Beckmann, K.B. and Ames, B.N. (1997). Oxidants, antioxidants, and aging. In: *Oxidative Stress and the Molecular Biology of Antioxidant Defenses* (Scandalios, J.G., Ed.). Cold Spring Harbor Laboratory Press, Cold Spring Harbor, NY, pp. 201–246.

Beckmann, R.P., Mizzen, L.A., and Welch, W.J. (1990). Interactions of hsp70 with newly synthesized proteins: implications for protein folding and assembly. *Science* 248: 850–854.

Beere, H.M. and Green, D.R. (2001). Stress management: heat shock protein-70 and the regulation of apoptosis. *Trends Cell Biol.* 11: 6–10.

Bell, J., Nelson, L., and Pellegrini, M. (1988). Effect of heat shock on ribosome synthesis in *Drosophila melanogaster. Mol. Cell. Biol.* 8: 91–95.

Bensaude, O., Pinto, M., Dubois, M.-F., Trung, N.V., and Morange, M. (1990). Protein denaturation during heat shock and related stress. In: *Stress Proteins: Induction and Function* (Schlesinger, M.J., Santoro, M.G., and Garaci, E., Eds.). Springer-Verlag, Heidelberg, pp. 89–99.

Bensaude, O., Bellier, S., Dubois, M.-F., Giannoni, F., and Nquyen, V.T. (1996). Heat shock induced protein modifications and modulation of enzyme activities. In: *Stress-Inducible Cellular Response* (Feige, U., Morimoto, R.I., Yahara, I., and Polla, B.S., Eds.). Birkhäuser, Basel, Switzerland, pp. 199–219.

Beyersmann, D. and Hechtenberg, S. (1997). Cadmium, gene regulation, and cellular signalling in mammalian cells. *Toxicol. Appl. Pharmacol.* 144: 247–261.

Black, A.R. and Subjeck, J.R. (1990). Mechanisms of stress-induced thermo- and chemotolerance. In: *Stress Proteins: Induction and Function* (Schlesinger, M.J., Santoro, M.G., and Garaci, E., Eds.). Springer-Verlag, Heidelberg, pp. 101–117.

Bohen, S.P. and Yamamoto, K.R. (1994). Modulation of steroid receptor signal transduction by heat shock proteins. In: *The Biology of Heat Shock Proteins and Molecular Chaperones* (Morimoto, R.I., Tissières, A., and Georgopoulos, C., Eds.) Cold Spring Harbor Laboratory Press, Cold Spring Harbor, NY, pp. 313–334.

Bollen, M. and Beullens, M. (2002). Signaling by protein phosphatases in the nucleus. *Trends Cell Biol.* 12: 138–145.

Bonneton, F., Theodore, L., Silar, P., Maroni, G., and Wegnez, M. (1996). Response of *Drosophila* metallothionein promoters to metallic, heat shock, and oxidative stresses. *FEBS Lett.* 380: 33–38.

Botella, J.A., Baines, I.A., Williams, D.D., Goberdhan, D.C.I., Proud, C.G., and Wilson, C. (2001). The *Drosophila* cell shape regulator c-Jun N-terminal kinase also functions as a stress-activated protein kinase. *Insect Biochem. Mol. Biol.* 31: 839–847.

Boyle, W.J., Smeal, T., Defize, L.H., Angel, P., Woodgett, J.R., Karin, M., and Hunter, T. (1991). Activation of protein kinase C decreases phosphorylation of c-Jun at sites that negatively regulate its DNA-binding activity. *Cell* 64: 573–584.

Brodsky, J.L. and Schekman, R. (1994). Heat shock cognate proteins and polypeptide translocation across the endoplasmic reticulum membrane. In: *The Biology of Heat Shock Proteins and Molecular Chaperones* (Morimoto, R.I., Tissières, A., and Georgopoulos, C., Eds.). Cold Spring Harbor Laboratory Press, Cold Spring Harbor, New York, pp. 85–109.

Broeks, A., Gerrard, B., Allikmets, R., Dean, M., and Plasterk, R.H. (1996). Homologues of the human multidrug resistance genes MRP and MDR contribute to heavy metal resistance in the soil nematode *Caenorhabditis elegans*. *EMBO J.* 15: 6132–6143.

Brostrom, C.O. and Brostrom, M.A. (1998). Regulation of transcriptional initiation during cellular responses to stress. *Prog. Nucl. Acid. Res. Mol. Biol.* 58: 79–125.

Brouwer, M. and Brouwer-Hoexum, T. (1998). Biochemical defense mechanisms against copper-induced oxidative damage in the blue crab, *Callinectes sapidus*. *Arch. Biochem. Biophys.* 351: 257–264.

Brun, A., Cuany, A., Le Mouel, T., Berge, J., and Amichot, M. (1996). Inducibility of the *Drosophila melanogaster* cytochrome P450 gene, CYP6A2, by phenobarbital in insecticide susceptible or resistant strains. *Insect Biochem. Mol. Biol.* 26: 697–703.

Burdon, R.H., Gill, V., and Rice-Evans, C.A. (1990). Active oxygen species and heat shock protein induction. In: *Stress Proteins: Induction and Function* (Schlesinger, M.J., Santoro, M.G., and Garaci, E., Eds.). Springer-Verlag, Heidelberg, pp. 19–25.

Burdon, R.H. (1994). Free radicals and cell proliferation. In: *Free Radical Damage and Its Control* (Rice-Evans, C.A. and Burdon, R.H., Eds.). Elsevier Science, Amsterdam, the Netherlands, pp. 155–185.

Burgering, B.M.T. and Bos, J.L. (1995). Regulation of Ras-mediated signalling: more than one way to skin a cat. *Trends Biochem. Sci.* 20: 18–22.

Burton, V., Mitchell, H.K., Young, P., and Petersen, N.S. (1988). Heat shock protection against cold stress of *Drosophila melanogaster*. *Mol. Cell. Biol.* 8: 3550–3552.

Cadenas, E. (1994). One- and two- electron activation of quinonoid compounds: oxidant and antioxidant aspects. In: *Free Radicals in the Environment, Medicine and Toxicology: Critical Aspects and Current Highlights* (Nohl, H., Esterbauer, H., and Rice-Evans, C.A., Eds.). Richelieu Press, London, pp. 119–135.

Calderwood, S.K. and Stevenson, M.A. (1993). Inducers of the heat shock response stimulate phospholipase C and phospholipase A activity in mammalian cells. *J. Cell. Physiol.* 155: 248–256.

Callaghan, A. and Denny, N. (2002). Evidence for an interaction between p-glycoprotein and cadmium toxicity in cadmium-resistant and -susceptible strains of *Drosophila melanogaster*. *Ecotox. Environ. Safety* 52: 211–213.

Calow, P. (1989). Proximate and ultimate responses to stress in biological systems. *Biol. J. Linn. Soc.* 37: 173–181.

Cavigelli, M., Li, W.W., Lin, A., Su, B., Yoshioka, K., and Karin, M. (1996). The tumor promoter arsenite stimulates AP-1 activity by inhibiting a JNK phosphatase. *EMBO J.* 15: 6269–6279.

Cerutti, P.A. (1985). Prooxidant states and tumor promotion. *Science* 227: 375–381.

Chaudère, J. (1994). Some chemical and biochemical constraints of oxidative stress in living cells. In: *Free Radical Damage and Its Control* (Rice-Evans, C.A. and Burdon, R.H., Eds.). Elsevier Science, Amsterdam, the Netherlands, pp. 25–66.

Chen, C.-P., Denlinger, D.L., and Lee, R.E., Jr. (1987). Cold-shock injury and rapid cold hardening in the flesh fly *Sarcophaga crassipalpis*. *Physiol. Zool.* 60: 297–304.

Chen, C.-P. and Walker, V.K. (1994). Cold-shock and chilling tolerance in *Drosophila*. *J. Insect Physiol.* 40: 661–669.

Chen, J.-H., Yu, C.-W., and Lin, L.-Y. (1999). Inactivation of metal-induced metallothionein gene expression by protein kinase C inhibitor. In: *Metallothionein IV* (Klaassen, C.D., Ed.). Birkhäuser, Basel, Switzerland, pp. 281–285.

Chen, J., Fujii, K., Zhang, L., Roberts, T., and Fu, H. (2001). Raf-1 promotes cell survival by antagonizing apoptosis signal-regulating kinase 1 through a MEK-ERK independent mechanism. *Proc. Natl. Acad. Sci. USA* 98: 7783–7788.

Chen, Q., Ma, E., Behar, K.L., Xu, T., and Haddad, G.G. (2002). Role of trehalose phosphate synthase in anoxia tolerance and development in *Drosophila melanogaster*. *J. Biol. Chem.* 277: 3274–3279.

Cherian, M.G. (1995). Metallothionein and its interaction with metals. In: *Toxicology of Metals* (Goyer, R.A. and Cherian, M.G., Eds.). Springer-Verlag, Berlin, pp. 121–138.

Chin, D. and Means, A.R. (2000). Calmodulin: a prototypical calcium sensor. *Trends Cell Biol.* 10: 322–328.

Clos, J., Rabindran, S., Wisniewski, J., and Wu, C. (1993). Induction temperature of human heat shock factor is reprogrammed in a *Drosophila* cell environment. *Nature* 364: 252–255.

Cornelius, G. (1996). Different stress factor signals converge at phosphoinositide turnover in *Drosophila* cells. *J. Therm. Biol.* 21: 85–89.

Cox, J.S., Shamu, C.E., and Walter, P. (1993). Transcriptional induction of genes encoding endoplasmic reticulum resident proteins requires a transmembrane protein kinase. *Cell* 73: 1197–1206.

Craig, E.A., Weissman, J.S., and Horwich, A.L. (1994). Heat shock proteins and molecular chaperones: mediators of protein conformation and turnover. *Cell* 78: 365–372.

Czajka, M.C. and Lee, R.E., Jr. (1990). A rapid cold-hardening response protecting against cold-shock injury in *Drosophila melanogaster*. *J. Exp. Biol.* 148: 245–254.

Dahlgaard, J., Loeschcke, V., Michalak, P., and Justesen, J. (1998). Induced thermotolerance and associated expression of the heat-shock protein hsp70 in adult *Drosophila melanogaster*. *Funct. Ecol.* 12: 786–793.

Dallinger, R. (1993). Strategies of metal detoxification in terrestrial invertebrates. In: *Ecotoxicology of Metals in Invertebrates* (Dallinger, R. and Rainbow, P.S., Eds.). Lewis Publishers, Boca Raton, FL, pp. 245–290.

Dallinger, R., Berger, B., Hunziker, R.E., and Kagi, J.H.R. (1997). Metallothionein in snail: Cd and Cu metabolism. *Nature* 388: 237–238.

Dalton, T.P., Li, Q.W., Bittel, D., Liang, L.C., and Andrews, G.K. (1996). Oxidative stress activates metal-responsive transcription factor-1 binding activity. Occupancy *in vivo* of metal responsive elements in the metallothionein-I gene promoter. *J. Biol. Chem.* 271: 26233–26241.

Dalton, T.P., Bittel, D., and Andrews, G.K. (1997). Reversible activation of mouse metal response element-binding transcription factor-1 DNA binding involves zinc interaction with the zinc finger domain. *Mol. Cell. Biol.* 17: 2781–2789.

DeMoor, J.M. and Koropatnick, D.J. (2000). Metals and cellular signaling in mammalian cells. *Cell. Mol. Biol.* (Noisy-le-Grand) 46: 367–381.

Denlinger, D.L., Joplin, K.H., Chen, C.-P., and Lee, R.E., Jr. (1991). Cold-shock and heat shock. In: *Insects at Low Temperature* (Lee, R.E., Jr. and Denlinger, D.L., Eds.). Chapman & Hall, New York, pp. 131–148.

Dérijard, B., Raingeaud, J., Barrett, T., Wu, I.-H., Han, J., Ulevitch, R.J., and Davis, R.J. (1995). Independent human MAP kinase signal transduction pathways defined by MEK and MKK isoforms. *Science* 267: 682–685.

Desagher, S. and Martinou, J.-C. (2000). Mitochondria as the central control point of apoptosis. *Trends Cell Biol.* 10: 369–377.

Deshpande, V.V. and Joshi, J.G. (1985). Vit. C Fe(III) induced loss of the covalently bound phosphate and enzyme activity of phosphoglucomutase. *J. Biol. Chem.* 260: 757–764.

Diplock, A.T. (1994). Antioxidants and free radical scavengers. In: *Free Radical Damage and Its Control* (Rice-Evans, C.A. and Burdon, R.H., Eds.). Elsevier Science, Amsterdam, the Netherlands, pp. 113–130.

Domenech, J., Palacios, O., Villarreal, L., González-Duarte, P., Capdevila, M., and Atrian, S. (2003). MTO: the second member of a *Drosophila* dual copper-thionein system. *FEBS Lett.* 533: 72–78.

Durliat, M., Bonneton, F., Boissonneau, E., André, M., and Wegnez, M. (1995). Expression of metallothionein genes during the post-embryonic development of *Drosophila melanogaster*. *BioMetals* 8: 339–351.

Duttaroy, A., Parkes, T., Emtage, P., Kirby, K., Boulianne, G.L., Wang, X., Hilliker, A.J., and Phillips, J.P. (1997). The manganese superoxide dismutase gene of *Drosophila*: structure, expression, and evidence for regulation by MAP kinase. DNA *Cell Biol.* 16: 391–399.

Elefant, F. and Palter, K.B. (1999). Tissue-specific expression of dominant negative mutant *Drosophila* HSC70 causes developmental defects and lethality. *Mol. Biol. Cell* 10: 2101–2117.

Ellis, R.J. and van der Vies, S.M. (1991). Molecular chaperones. *Annu. Rev. Biochem.* 60: 321–347.

Evan, G. and Littlewood, T. (1998). A matter of life and death. *Science* 281: 1317–1322.

Fabisiak, J.P., Tyurin, V.A., Tyurina, Y.Y., Borisenko, G.G., Korotaeva, A., Pitt, B.R., Lazo, J.S., and Kagan, V.E. (1999). Redox regulation of copper-metallothionein. *Arch. Biochem. Biophys.* 363: 171–181.

Feder, J.H., Rossi, J.M., Solomon, J., Solomon, N., and Lindquist, S. (1992). The consequences of expressing hsp70 in *Drosophila* cells at normal temperatures. *Genes Dev.* 6: 1402–1413.

Feder, M.E. (1996). Ecological and evolutionary physiology of stress proteins and the stress response: the *Drosophila melanogaster* model. In: *Animals and Temperature: Phenotypic and Evolutionary Adaptation* (Johnston, I.A. and Bennett, A.F., Eds.). Cambridge University Press, Cambridge, pp. 79–102.

Feder, M.E., Cartaño, N.V., Milos, L., Krebs, R.A., and Lindquist, S. (1996). Effect of engineering hsp70 copy number on hsp70 expression and tolerance of ecologically relevant heat shock in larvae and pupae of *Drosophila melanogaster*. *J. Exp. Biol.* 199: 1837–1844.

Feder, M.E., Blair, N., and Figueras, H. (1997). Natural thermal stress and heat shock protein expression in *Drosophila* larvae and pupae. *Funct. Ecol.* 11: 90–100.

Feder, M.E. and Hofmann, G.E. (1999). Heat-shock proteins, molecular chaperones, and the stress response: evolutionary and ecological physiology. *Annu. Rev. Physiol.* 61: 243–282.

Feder, M.E., Bennett, A.F., and Huey, R.B. (2000). Evolutionary physiology. *Annu. Rev. Ecol. Syst.* 31: 315–341.

Fernandes, M., O'Brien, T., and Lis, J.T. (1994). Structure and regulation of heat shock gene promoters. In: *The Biology of Heat Shock Proteins and Molecular Chaperones* (Morimoto, R.I., Tissières, A., and Georgopoulos, C., Eds.). Cold Spring Harbor Laboratory Press, Cold Spring Harbor, New York, pp. 375–393.

Feyereisen, R. (1999). Insect P450 enzymes. *Annu. Rev. Entomol.* 44: 507–533.

Flores, S.C. and McCord, J.M. (1997). Redox regulation by the HIV-1 Tat transcriptional factor. In: *Oxidative Stress and the Molecular Biology of Antioxidant Defenses* (Scandalios, J.G., Ed.). Cold Spring Harbor Laboratory Press, Cold Spring Harbor, New York, pp. 117–138.

Fogleman, J.C., Danielson, P.B., and Macintyre, R.J. (1998). The molecular basis of adaptation in *Drosophila*: the role of cytochrome P450s. In: *Evolutionary Biology* (Hecht, M.K., Macintyre, R.J., and Clegg, M.I., Eds.). Plenum Press, New York, pp. 15–77.

Freedman, J.H., Slice, L.W., Dixon, D., Fire, A., and Rubin, C.S. (1993). The novel metallothionein genes of *Caenorhabditis elegans*: structural organisation and inducible, cell-specific expression. *J. Biol. Chem.* 268: 2554–2564.

Fuchs, D., Baier-Bitterlich, G., Wede, I., and Wachter, H. (1997). Reactive oxygen and apoptosis. In: *Oxidative Stress and the Molecular Biology of Antioxidant Defenses* (Scandalios, J.G., Ed.). Cold Spring Harbor Laboratory Press, Cold Spring Harbor, New York, pp. 139–167.

Fuchs, S.Y., Spiegelman, V.S., and Belitsky, G.A. (1994). Inducibility of various cytochrome P450 isozymes by phenobarbital and some other xenobiotics in *Drosophila melanogaster*. *Biochem. Pharmacol.* 47: 1867–1873.

Gardner, A.M., Xu, F.-H., Fady, C., Jacoby, F.J., Duffy, D.C., Tu, Y., and Lightenstein, A. (1997). Apoptotic vs nonapoptotic cytotoxicity induced by hydrogen peroxide. *Free Rad. Biol. Med.* 22: 73–83.

Gems, D. (1999). Nematode ageing: putting metabolic theories to the test. *Curr. Biol.* 9: R614–R616.

Gems, D. and Partridge, L. (2001). Insulin/IGF signalling and ageing: seeing the bigger picture. *Curr. Opin. Genet. Dev.* 11: 287–292.

Georgieva, T.G., Jankulova, E.D., Ralchev, K.H., and Dunkov, B.C. (1995). Induction of diaphorase-1 by dicoumarol in *Drosophila virilis* larvae. *Arch. Insect. Biochem. Physiol.* 29: 25–34.

Gething, M.-J. and Sambrook, J.F. (1992). Protein folding in the cell. *Nature* 355: 33–45.

Gething, M.-J., Blond-Elguindi, S., Mori, K., and Sambrook, J.F. (1994). Structure, function, and regulation of the endoplasmic reticulum chaperone, BiP. In: *The Biology of Heat Shock Proteins and Molecular Chaperones* (Morimoto, R.I., Tissières, A., and Georgopoulos, C., Eds.). Cold Spring Harbor Laboratory Press, Cold Spring Harbor, New York, pp. 111–135.

Giedroc, D.P., Chen, X., and Apuy, J.L. (2001). Metal response element (MRE)-binding trnascription factor-1 (MTF-1): structure, function, and regulation. *Antiox. Redox Sign.* 3: 577–596.

Gill, H.J., Nida, D.L., Dean, D.A., England, M.W., and Jacobson, J.B. (1989). Resistance of *Drosophila* to cadmium: biochemical factors in resistant and sensitive strains. *Toxicology* 56: 315–321.

Gill, J.H. and Dive, C. (2000). Apoptosis: basic mechanisms and relevance to toxicology. In: *Apoptosis in Toxicology* (Roberts, R., Ed.). Taylor & Francis, London, pp. 1–19.

Goeptar, A.R. (1993). The role of cytochrome P450 in the reductive bioactivation of cytostatic quinones: a molecular toxicological study. Ph.D. thesis, Vrije Universiteit, Amsterdam, the Netherlands.

Goering, P.L., Fisher, B.R., and Kish, C.L. (1993). Stress protein synthesis induced in rat liver by cadmium precedes hepatotoxicity. *Toxicol. Appl. Pharmacol.* 122: 139–148.

Goering, P.L. and Fisher, B.R. (1995). Metals and stress proteins. In: *Toxicology of Metals* (Goyer, R.A. and Cherian, M.G., Eds.). Springer-Verlag, Berlin, pp. 229–266.

Goering, P.L., Waalkes, M.P., and Klaassen, C.D. (1995). Toxicology of cadmium. In: *Toxicology of Metals* (Goyer, R.A. and Cherian, M.G., Eds.). Springer-Verlag, Berlin, pp. 189–214.

Gonzalez, F.J., Liu, S.-Y., and Yano, M. (1993). Regulation of cytochrome P450 genes: molecular mechanisms. *Pharmacogenetics* 3: 51–57.

Goossens, V., Grooten, J., De Vos, K., and Fiers, W. (1995). Direct evidence for tumor necrosis factor-induced mitochondrial reactive oxygen intermediates and their involvement in cytotoxicity. *Proc. Natl. Acad. Sci. USA* 92: 8115–8119.

Gottifredi, V., Shieh, S.Y., and Prive, C. (2000). Regulation of p53 after different forms of stress and at different cell cycle stages. In: *Cold Spring Harbor Symposium on Quantitative Biology: Biological Responses to DNA Damage.* Cold Spring Harbor Laboratory Press, Cold Spring Harbor, New York, Vol. 65, pp. 483–488.

Green, D.R. and Reed, J.C. (1998). Mitochondria and apoptosis. *Science* 281: 1309–1312.

Gutenby, A.A., Donaldson, G.K., Coloubinoff, P., LaRossa, R.A., Lorimer, G.H., Lubben, T.H., van Dijk, T.K., and Viitanen, P.V. (1990). The cellular functions of chaperones. In: *Stress Proteins: Induction and Function* (Schlesinger, M.J., Santoro, M.G., and Garaci, E., Eds.). Springer-Verlag, Heidelberg, pp. 57–69.

Guyton, K.Z., Liu, Y., Gorospe, M., Xu, Q., and Holbrook, N.J. (1996). Activation of mitogen-activated protein kinase by H_2O_2. *J. Biol. Chem.* 271: 4138–4142.

Guyton, K.Z., Gorospe, M., and Holbrook, N.J. (1997). Oxidative stress, gene expression, and the aging process. In: *Oxidative Stress and the Molecular Biology of Antioxidant Defenses* (Scandalios, J.G., Ed.). Cold Spring Harbor Laboratory Press, Cold Spring Harbor, New York, pp. 247–272.

Hahn, G.M. and Li, G.C. (1990). Thermotolerance, thermoresistance, and thermosensitization. In: *Stress Proteins in Biology and Medicine* (Morimoto, R.I., Tissières, A., and Georgopoulos, C., Eds.). Cold Spring Harbor Laboratory Press, Cold Spring Harbor, New York, pp. 79–100.

Hamer, D.H. (1986). Metallothionein. *Annu. Rev. Biochem.* 55: 931–951.

Han, S.-J., Choi, K.-Y., Brey, P.T., and Lee, W.-J. (1998). Molecular cloning and characterization of a *Drosophila* p38 mitogen-activated protein kinase. *J. Biol. Chem.* 273: 369–374.

Hannun, Y.A. (1996). Functions of ceramide in coordinating cellular responses to stress. *Science* 274: 1855–1859.

Hannun, Y.A. and Luberto, C. (2000). Ceramide in the eukaryotic stress response. *Trends Cell Biol.* 10: 73–80.

Harshman, L.G. and James, A.A. (1998). Differential gene expression in insects: transcriptional control. *Annu. Rev. Entomol.* 43: 671–700.

Hartl, F.U. (1996). Molecular chaperones in cellular protein folding. *Nature* 381: 571–580.

Hauptmann, N. and Cadenas, E. (1997). The oxygen paradox: biochemistry of active oxygen. In: *Oxidative Stress and the Molecular Biology of Antioxidant Defenses* (Scandalios, J.G., Ed.). Cold Spring Harbor Laboratory Press, Cold Spring Harbor, New York, pp. 1–20.

Hensbergen, P.J. (1999). Metallothionein in *Orchesella cincta*. Ph.D. thesis, Vrije Universiteit, Amsterdam, the Netherlands.

Hightower, L.E., Sadis, S.E., and Takenaka, I.M. (1994). Interactions of vertebrate hsc70 and hsp70 with unfolded proteins and peptides. In: *The Biology of Heat Shock Proteins and Molecular Chaperones* (Morimoto, R.I., Tissières, A., and Georgopoulos, C., Eds.). Cold Spring Harbor Laboratory Press, Cold Spring Harbor, New York, pp. 179–207.

Hill, C.S. and Treisman, R. (1995). Transcriptional regulation by extracellular signals: mechanisms and specificity. *Cell* 80: 199–211.

Hoffmann, A.A. and Parsons, P.A. (1991). *Evolutionary Genetics and Environmental Stress*. Oxford University Press, Oxford.

Holbrook, N.J., Liu, Y., and Fornace, A.J., Jr. (1996). Signaling events controlling molecular response to genotoxic stress. In: *Stress-Inducible Cellular Response* (Feige, U., Morimoto, R.I., Yahara, I., and Polla, B.S., Eds.). Birkhäuser, Basel, Switzerland, pp. 273–288.

Holzenberger, M., Dupont, J., Ducos, B., Leneuve, P., Géloën, A., Evens, P.C., Cervera, P., and Le Bouc, Y. (2003). IGF-1 receptor regulates lifespan and resistance to oxidative stress in mice. *Nature* 421: 182–187.

Hopkin, S.P. (1989). *Ecophysiology of Metals in Terrestrial Invertebrates*. Elsevier Applied Sciences, London.

Huang, L.E., Zhang, H., Bae, S.W., and Liu, A.Y.-C. (1994). Thiol reducing reagents inhibit the heat shock response (involvement of a redox mechanism in the heat shock signal transduction pathway). *J. Biol. Chem.* 269: 30718–30725.

Huey, R.B. and Bennett, A.F. (1990). Physiological adjustments to fluctuating thermal environments: an ecological and evolutionary perspective. In: *Stress Proteins in Biology and Medicine* (Morimoto, R.I., Tissières, A., and Georgopoulos, C., Eds.). Cold Spring Harbor Laboratory Press, Cold Spring Harbor, New York, pp. 37–59.

Hunter, T. and Karin, M. (1992). The regulation of transcription by phosphorylation. *Cell* 70: 375–387.

Ivessa, N.S. (2002). Assembling new pieces of the UPR puzzle. *Trends Cell Biol.* 12: 109.

Jabs, T. (1999). Reactive oxygen intermediates as mediators of programmed cell death in plants and animals. *Biochem. Pharmacol.* 57: 231–245.

Jacquier-Sarlin, M.R., Jornot, L., and Polla, B.S. (1995). Differential expression and regulation of hsp70 and hsp90 by phorbol esters and heat shock. *J. Biol. Chem.* 270: 14094–14099.

Jacquier-Sarlin, M.R. and Polla, B.S. (1996). Dual regulation of heat-shock transcription factor (HSF) activation and DNA-binding activity by H_2O_2: role of thioredoxin. *Biochem. J.* 318: 187–193.

Jaiswal, A.K. (1994). Antioxidant response element. *Biochem. Pharmacol.* 48: 139–141.

Jamieson, D.J. and Storz, G. (1997). Transcriptional regulators of oxidative stress responses. In: *Oxidative Stress and the Molecular Biology of Antioxidant Defenses* (Scandalios, J.G., Ed.). Cold Spring Harbor Laboratory Press, Cold Spring Harbor, New York, pp. 91–115.

Jedlicka, P., Mortin, M.A., and Wu, C. (1997). Multiple functions of *Drosophila* heat shock transcription factor in vivo. *EMBO J.* 16: 2452–2462.

Jiang, L.-J., Maret, W., and Vallee, B.L. (1998). The glutathione redox couple modulates zinc transfer from metallothionein to zinc-depleted sorbitol dehydrogenase. *Proc. Natl. Acad. Sci. USA* 95: 3483–3488.

Jiménez, I., Aracena, P., Letelier, M.E., Navarro, P., and Speisky, H. (2002). Chronic exposure of HepG2 cells to excess copper results in depletion of glutathione and induction of metallothionein. *Toxicol. in vitro* 16: 167–175.

Karin, M. (1985). Metallothioneins: proteins in search of function. *Cell*. 41: 9–10.

Karin, M. (1995). The regulation of AP-1 activity by mitogen-activated protein kinases. *J. Biol. Chem.* 270: 16483–16486.

Kasai, S. and Scott, J.G. (2000). Overexpression of cytochrome P450 CYP6D1 is associated with monooxygenase-mediated pyrethroid resistance in house flies from Georgia. *Pestic. Biochem. Physiol.* 68: 34–41.

Kassenbrock, C.K., Garcia, P.D., Walter, P., and Kelly, R.B. (1988). Heavy chain binding protein recognizes aberrant polypeptides translated in vitro. *Nature* 333: 90–93.

Kaufmann, S.H. and Hengartner, M.O. (2001). Programmed cell death: alive and well in the new millennium. *Trends Cell Biol.* 12: 526–534.

Kaul, N. and Forman, H.J. (2000). Reactive oxygen species in physiology and toxicology. In: *Toxicology of the Human Environment: The Critical Role of Free Radicals* (Rhodes, C.J., Ed.). Taylor & Francis, London, pp. 311–335.

Kawanishi, S. (1995). Role of active oxygen species in metal-induced DNA damage. In: *Toxicology of Metals* (Goyer, R.A. and Cherian, M.G., Eds.). Springer-Verlag, Berlin, pp. 349–372.

Keeton, W.T. and Gould, J.L. (1993). *Biological Science* (5th ed.). Norton, New York.

Kelley, P.M. and Schlesinger, M.J. (1978). The effects of amino acid analogs and heat shock in gene expression in chicken embryo fibroblasts. *Cell* 19: 1277–1286.

Kelty, J.D. and Lee, R.E., Jr. (1999). Induction of rapid cold hardening by cooling at ecologically relevant rates in *Drosophila melanogaster*. *J. Insect Physiol.* 45: 719–726.

Kelty, J.D. and Lee, R.E., Jr. (2001). Rapid cold-hardening of *Drosophila melanogaster* (Diptera: Drosophilidae) during ecologically based thermoperiodic cycles. *J. Exp. Biol.* 204: 1659–1666.

Kenyon, C., Chang, J., Gensch, E., Rudner, A. and Tabtlang, R. (1993). A *C. elegans* mutant that lives twice as long as wild type. *Nature* 366: 461–464.

Kholodenko, B.N. (2002). MAP kinase cascade signaling and endocytic trafficking: a marriage of convenience? *Trends Cell Biol.* 12: 173–177.

Kimura, M.T., Awasaki, T., Ohtsu, T., and Shimada, K. (1992). Seasonal changes in glycogen and trehalose content in relation to winter survival of four temperate species of *Drosophila*. *J. Insect Physiol.* 38: 871–875.

Kimura, K.D., Tissenbaum, H.A., Liu, Y., and Ruvkun, G. (1997). daf-2, an insulin receptor-like gene that regulates longevity and diapause in *Caenorhabditis elegans*. *Science* 277: 942–946.

Kleiber, M., Ed. (1961). *The Fire of Life: An Introduction to Animal Energetics*. John Wiley & Sons, New York.

Kohno, K., Normington, K., Sambrook, J., Gething, M.-J., and Mori, K. (1993). The promoter region of the yeast KAR2 (BiP) gene contains a regulatory domain that responds to the presence of unfolded proteins in the endoplasmic reticulum. *Mol. Cell. Biol.* 13: 877–890.

Koppenol, W.H. (1994). Chemistry of iron and copper in radical reactions. In: *Free Radical Damage and Its Control* (Rice-Evans, C.A. and Burdon, R.H., Eds.). Elsevier Science, Amsterdam, the Netherlands, pp. 3–24.

Koropatnick, J. and Leibbrandt, M.E.I. (1995). Effects of metals on gene expression. In: *Toxicology of Metals* (Goyer, R.A. and Cherian, M.G., Eds.). Springer-Verlag, Berlin, pp. 93–120.

Koster, J.F. and Sluiter, W. (1994). Physiological relevance of free radicals and their relation to iron. In: *Free Radicals in the Environment, Medicine, and Toxicology: Critical Aspects and Current Highlights* (Nohl, H., Esterbauer, H., and Rice-Evans, C.A., Eds.). Richelieu Press, London, pp. 409–427.

Kowald, A. and Kirkwood, T.B.L. (1994). Towards a network theory of ageing: a model combining the free radical theory and the protein error theory. *J. Theor. Biol.* 168: 75–94.

Krauss, G. (2001). Apoptosis. In: *Biochemistry of Signal Transduction and Regulation* (Krauss, G., Ed.) (2nd ed.). Wiley-VCH., Weinheim, Germany, pp. 455–472.

Krebs, R.A. and Loeschcke, V. (1994). Costs and benefits of activation of the heat-shock response in *Drosophila melanogaster*. *Funct. Ecol.* 8: 730–737.

Krebs, R.A. and Feder, M.E. (1997a). Tissue-specific variation in hsp70 expression and thermal damage in *Drosophila melanogaster* larvae. *J. Exp. Biol.* 200: 2007–2015.

Krebs, R.A. and Feder, M.E. (1997b). Deleterious consequences of hsp70 overexpression in *Drosophila melanogaster* larvae. *Cell Stress Chaperones* 2: 60–71.

Krebs, R.A. and Feder, M.E. (1997c). Natural variation in the expression of the heat-shock protein hsp70 in a population of *Drosophila melanogaster*, and its correlation with tolerance of ecological relevant thermal stress. *Evolution* 51: 173–179.

Krebs, R.A. and Feder, M.E. (1998). Hsp70 and larval thermotolerance in *Drosophila melanogaster*. How much is enough and when is more too much? *J. Insect. Physiol.* 44: 1091–1101.

Krishnan, V. and Safe, S. (1993). Polychlorinated biphenyls (PCBs), dibenzo-p-dioxins (PCDDs), and dibenzofurans (PCDFs) as antiestrogens in MCG-7 human breast cancer cells: quantitative structure-activity relationships. *Toxicol. Appl. Pharmacol.* 120: 55–61.

Kuether, K. and Arking, R. (1999). *Drosophila* selected for extended longevity are more sensitive to heat shock. *Age* (Media) 22: 175–180.

Kunau, W.-H., Agne, B., and Girzalsky, W. (2001). The diversity of organelle protein transport mechanisms. *Trends Cell Biol.* 11: 358–361.

Kyriakis, J.M. and Avruch, J. (1996). Sounding the alarm: protein kinase cascades activated by stress and inflamnation. *J. Biol. Chem.* 271: 24313–24316.

Lafont, R. and Connat, J.L. (1989). Pathways of ecdysone metabolism. In: *Ecdysone: From Chemistry to Mode of Action* (Koolman, J., Ed.). Thieme Verlag, Stuttgart, Germany, pp. 167–173.

Landers, J.P. and Bunce, N.J. (1991). The Ah-receptor and the mechanism of dioxin toxicity. *Biochem. J.* 276: 273–287.

Langer, T. and Neupert, W. (1994). Chaperoning mitochondrial biogenesis. In: *The Biology of Heat Shock Proteins and Molecular Chaperones* (Morimoto, R.I., Tissières, A., and Georgopoulos, C., Eds.). Cold Spring Harbor Laboratory Press, Cold Spring Harbor, New York, pp. 53–83.

Laszlo, A. (1992). The thermoresistant state: protection from initial damage or better repair? *Exp. Cell. Res.* 202: 519–531.

Latchman, D.S. (1995). *Eukaryotic Transcription Factors*, 2nd ed., Academic Press, San Diego, CA.

Lauverjat, S., Ballan-Dufrancais, C., and Wegnez, M. (1989). Detoxification of cadmium: ultrastructural study and electron-probe microanalysis of the midgut in a cadmium resistant strain of *Drosophila melanogaster. Biol. Metals* 2: 97–107.

Lavy, D. (1996). Nutritional ecology of soil arthropods. Ph.D. thesis, Vrije Universiteit, Amsterdam, the Netherlands.

Lee, R.E., Jr. (1991). Principles of insects' low temperature tolerance. In: *Insects at Low Temperature* (Lee, R.E., Jr. and Denlinger, D.L., Eds.). Chapman & Hall, New York, pp. 17–46.

Lee, R.E., Jr. and Denlinger, D.L., Eds. (1991). *Insects at Low Temperature.* Chapman & Hall, New York,.

Lee, Y.J. and Shacter, E. (1999). Oxidative stress inhibits apoptosis in human lymphoma cells. *J. Biol. Chem.* 274: 19792–19798.

Leist, M., Single, B., Castoldi, A., Kuhne, S., and Nicotera, P. (1997). Intracellular adenosine triphosphate (ATP) concentration: a switch in the decision between apoptosis and necrosis. *J. Exp. Med.* 185: 1481–1486.

Li, X. and Lee, A.S. (1991). Competitive inhibition of a set of endoplasmic reticulum protein genes (GRP78, GRP94 and Erp72) retards cell growth and lowers viability after ionophore treatment. *Mol. Cell. Biol.* 11: 3446–3453.

Liao, V.H.-C., Dong, J., and Freedman, J.H. (2002). Molecular characterization of a novel, cadmium-inducible gene from the nematode *Caenorhabditis elegans. J. Biol. Chem.* 277: 42049–42059.

Lin, Y.J., Seroude, L., and Benzer, S. (1998). Extended life-span and stress resistance in the *Drosophila* mutant *methuselah. Science* 282: 943–946.

Lindquist, S. (1986). The heat-shock response. *Annu. Rev. Biochem.* 55: 1151–1191.

Lindquist, S. and Craig, E.A. (1988). The heat-shock proteins. *Annu. Rev. Genet.* 55: 631–677.

Lis, J.T., Xiao, H., and Perisic, O. (1990). Modular units of heat shock regulatory regions: structures and functions. In: *Stress Proteins in Biology and Medicine* (Morimoto, R.I., Tissières, A., and Georgopoulos, C., Eds.). Cold Spring Harbor Laboratory Press, Cold Spring Harbor, New York, pp. 411–428.

Lis, J.T. and Wu, C. (1993). Protein traffic on the heat shock promoter: parking, stalling, and trucking along. *Cell* 74: 1–4.

Liu, N. and Scott, J.G. (1998). Increased transcription of CYP6D1 causes cytochrome P450-mediated insecticide resistance in house fly. *Insect Biochem. Mol. Biol.* 28: 531–535.

Loeschcke, V., Krebs, R.A., Dahlgaard, J., and Michalak, P. (1997). High-temperature stress and the evolution of thermal resistance in *Drosophila*. In: *Environmental Stress, Adaptation, and Evolution* (Bijlsma, R. and Loeschcke, V., Eds.). Birkhäuser, Basel, Switzerland, pp. 175–190.

Lohrum, M.A.E. and Vousden, K.H. (2000). Regulation and function of the p53-related proteins: same family, different rules. *Trends Cell Biol.* 10: 197–202.

Lundberg, A.S., Hahn, W.C., Gupta, P., and Weinberg, R.A. (2000). Genes involved in senescence and immortalization. *Curr. Opin. Cell Biol.* 12: 705–709.

Luz, J.M. and Lennarz, W.J. (1996). Protein disulfide isomerase: a multi-functional protein of the endoplasmic reticulum. In: *Stress-Inducible Cellular Responses* (Feige, U., Morimoto, R.I., Yahara, I., Polla, B.S., Eds.). Birkhäuser, Basel, Switzerland, pp. 97–117.

Mager, W.H. and De Kruijff, A.J.J. (1995). Stress-induced transcriptional activation. *Microbiol. Rev.* 59: 506–531.

Maltby, L. (1999). Studying stress: the importance of organism-level responses. *Ecol. Appl.* 9: 431–440.

Maroni, G., Wise, J., Young, J.E., and Otto, E. (1987). Metallothionein gene duplications and metal tolerance in natural populations of *Drosophila melanogaster*. *Genetics* 117: 739–744.

Maroni, G., Ho, A.-S., and Theodore, L. (1995). Genetic control of cadmium tolerance in *Drosophila melanogaster*. *Environ. Health Perspect.* 103: 116–118.

Marshall, C.J. (1995). Specificity of receptor tyrosine signaling: transient versus sustained extracellular signal-regulated kinase activity. *Cell* 80: 179–185.

Martin, G.M., Austad, S.N., and Johnson, T.E. (1996). Genetic analysis of ageing: role of oxidative damage and environmental stresses. *Nature Genet.* 13: 25–34.

Martin-Blanco, E. (2000). MAPK signalling cascades: ancient roles and new functions. *BioEssays* 22: 637–645.

McCord, J.M. (1995). Superoxide radical: controversies, contradictions, and paradoxes. *Proc. Soc. Exp. Biol. Med.* 209: 112–117.

McKay, D.B., Wilbanks, S.M., Flaherty, K.M., Ha, J.-H., O'Brien, M.C., and Shirvanee, L.L. (1994). Stress-70 proteins and their interaction with nucleotides. In: *The Biology of Heat Shock Proteins and Molecular Chaperones* (Morimoto, R.I., Tissières, A., and Georgopoulos, C., Eds.). Cold Spring Harbor Laboratory Press, Cold Spring Harbor, New York, pp. 153–177.

Mehlen, P., Kretz-Remy, C., Preville, X., and Arrigo, A.-P. (1996a). Human hsp27, *Drosophila* hsp27, and human alpha-B-crystallin expression-mediated increase in glutathione is essential for the protective activity of these proteins against TNF-alpha-induced cell death. *EMBO J.* 15: 2695–2706.

Mehlen, P., Schulzeosthoff, K., and Arrigo, A.-P. (1996b). Small stress proteins as novel regulators of apoptosis: heat shock protein 27 blocks Fas/APO-1- and staurosporine-induced cell death. *J. Biol. Chem.* 271: 16510–16514.

Mehlen, P., Hickey, E., Weber, L.A., and Arrigo, A.-P. (1997). Large unphosphorylated aggregates as the active form of hsp27 which controls intracellular reactive oxygen species and glutathione levels and generates a protection against TNFα in NIH 3T3-ras cells. *Biochem. Biophys. Res. Commun.* 241: 187–192.

Min, K.-S., Nishida, K., Nakahara, Y., and Onosaka, S. (1999). Protective effect of metallothionein on DNA damage induced by hydrogen peroxide and ferric iron-nitrilotriacetic acid. In: *Metallothionein IV* (Klaassen, C.D., Ed.). Birkhäuser, Basel, Switzerland, pp. 529–534.

Misener, S.R., Chen, C.-P., and Walker, V.K. (2001). Cold tolerance and proline metabolic gene expression in *Drosophila melanogaster*. *J. Insect Physiol.* 47: 393–400.

Mizzen, L.A. and Welch, W.J. (1988). Characterization of the thermotolerant cell. I. Effects on protein synthesis activity and the regulation of heat shock protein 70 expression. *J. Cell. Biol.* 106: 1105–1116.

Mockett, R.J., Orr, W.C., Rahmandar, J.J., Benes, J.J., Radyuk, S.N., Klichko, V.I., and Sohal, R.S. (1999a). Overexpression of Mn-containing superoxide dismutase in transgenic *Drosophila melanogaster*. *Arch. Biochem. Biophys.* 37: 260–269.

Mockett, R.J., Sohal, R.S., and Orr, W.C. (1999b). Overexpression of glutathione reductase extends survival in transgenic *Drosophila melanogaster* under hypoxia but not under normoxia. *FASEB J.* 13: 1733–1742.

Mockett, R.J., Bayne, A.-C.V., Kwong, L.K., Orr, W.C., and Sohal, R.S. (2003). Ectopic expression of catalase in *Drosophila* mitochondria increases stress resistance but not longevity. *Free Rad. Biol. Med.* 34: 207–217.

Mori, K., Ma, W., Gething, M.-J., and Sambrook, J. (1993). A transmembrane protein with a cdc^{2+}/CDC28-related kinase activity is required for signaling from the ER to the nucleus. *Cell* 74: 743–756.

Morimoto, R.I. and Milarski, K.L. (1990). Expression and function of vertebrate hsp70 genes. In: *Stress Proteins in Biology and Medicine* (Morimoto, R.I., Tissières, A., and Georgopoulos, C., Eds.). Cold Spring Harbor Laboratory Press, Cold Spring Harbor, New York, pp. 323–359.

Morimoto, R.I., Abravaya, K., Mosser, D., and Williams, G.T. (1990a). Transcription of the human hsp70 gene: Cis-acting elements and transacting factors involved in basal, adenovirus E1A, and stress-induced expression. In: *Stress Proteins: Induction and Function* (Schlesinger, M.J., Santoro, M.G., and Garaci, E., Eds.). Springer-Verlag, Heidelberg, pp. 1–18.

Morimoto, R.I., Tissières, A., and Georgopoulos, C. (1990b). The stress response, function of the proteins, and perspectives. In: *Stress Proteins in Biology and Medicine* (Morimoto, R.I., Tissières, A., and Georgopoulos, C., Eds.). Cold Spring Harbor Laboratory Press, Cold Spring Harbor, New York, pp. 1–36.

Morimoto, R.I. (1993). Cells in stress: transcriptional activation of heat shock genes. *Science* 259: 1409–1410.

Morimoto, R.I., Tissières, A., and Georgopoulos, C. (1994a). Progress and perspectives on the biology of heat shock proteins and molecular chaperones. In: *The Biology of Heat Shock Proteins and Molecular Chaperones* (Morimoto, R.I., Tissières, A., and Georgopoulos, C. Eds.). Cold Spring Harbor Laboratory Press, Cold Spring Harbor, New York, pp. 1–30.

Morimoto, R.I., Jurivich, D.A., Kroeger, P.E., Mathur, S.K., Murphy, S.P., Nakai, A., Sarge, K., Abravaya, K., and Sistonen, L.T. (1994b). Regulation of heat shock proteins and molecular chaperones. In: *The Biology of Heat Shock Proteins and Molecular Chaperones*. (Morimoto, R.I., Tissières, A., and Georgopoulos, C., Eds.). Cold Spring Harbor Laboratory Press, Cold Spring Harbor, New York, pp. 417–455.

Mosser, D.D., Duchaine, J., and Massie, B. (1993). The DNA-binding activity of human heat shock transcription factor is regulated *in vivo* by hsp70. *Mol. Cell. Biol.* 13: 5427–5438.

Mosser, D.D., Caron, A.W., Bourget, L., Dennis-Larose, C., and Massie, B. (1997). Role of the human heat shock protein hsp70 in protection against stress-induced apoptosis. *Mol. Cell. Biol.* 17: 5317–5327.

Munks, R.J.L. and Turner, B.M. (1994). Suppression of heat-shock protein synthesis by short-chain fatty acids and alcohols. *Biochim. Biophys. Acta* 1223: 23–28.

Nebert, D.W., Adesink, M., Coon, M.J., Estabrook, R.W., and Gonzalez, F.J. (1987). The P450 gene superfamily: recommended nomenclature. *DNA* 6: 1–11.

Nebert, D.W., Roe, A.L., Dieter, M.Z., Solis, W.A., Yang, Y., and Dalton, T.P. (2000). Role of the aromatic hydrocarbon receptor and (Ah) gene battery in the oxidative stress response, cell cycle control, and apoptosis. *Biochem. Pharmacol.* 59: 65–85.

Nelson, R.J., Ziegelhoffer, T., Nicolet, C., Werner-Washburner, M., and Craig, E.A. (1992). The translation machinery and the 70 kD heat-shock protein cooperate in protein synthesis. *Cell* 71: 97–105.

Nelson, S.K., Bose, S.K., and McCord, J.M. (1994). The toxicity of high-dose superoxide dismutase suggests that superoxide can both initiate and terminate lipid peroxidation in the reperfused heart. *Free Rad. Biol. Med.* 16: 195–200.

Ohtsu, T., Kimura, M.T., and Katagiri, C. (1998). How *Drosophila* species acquire cold tolerance: qualitative changes of phospholipids. *Eur. J. Biochem.* 252: 608–611.

Oikawa, S., Kurasaki, M., Kojima, Y., and Kawanishi, S. (1995). Oxidative and nonoxidative mechanisms of site-specific DNA cleavage induced by copper-containing metallothioneins. *Biochemistry* 34: 8763–8770.

Oliver, C.N., Ann, B.-W., Moerman, E.J., Goldskin, S., and Stadtman, E.R. (1987). Age-related changes in oxidized proteins. *J. Biol. Chem.* 262: 5488–5492.

Orr, W.C. and Sohal, R.S. (1994). Extension of life-span by overexpression of superoxide dismutase and catalase in *Drosophila melanogaster*. *Science* 263: 1128–1130.

Palleros, D.R., Reid, K.L., Shi, L., Wel, W.J., and Fink, A.L. (1993). ATP-induced protein-hsp70 complex dissociation requires K+ but not ATP hydrolysis. *Nature* 365: 664–666.

Palmiter, R.D. (1994). Regulation of metallothionein gene by heavy metals appears to be mediated by a zinc sensitive inhibitor that interacts with a constitutive active transcription factor, MTF-1. *Proc. Natl. Acad. Sci.*, USA, 91: 1219–1223.

Palmiter, R.D. (1998). The elusive function of metallothioneins. *Proc. Natl. Acad. Sci. USA* 95: 8428–8430.

Parker, E.D., Jr., Forbes, V.E., Nielsen, S.L., Ritter, C., Barata, C., Baird, D.J., Admiraal, W., Levin, L., Loeschcke, V., Lyytikäinen-Saarenmaa, P., Høgh-Jensen, H., Calow P., and Ripley, B.J. (1999). Stress in ecological systems. *Oikos* 86: 179–184.

Parsell, D.A. and Lindquist, S. (1994). Heat shock proteins and stress tolerance. In: *The Biology of Heat Shock Proteins and Molecular Chaperones* (Morimoto, R.I., Tissières, A., and Georgopoulos, C., Eds.). Cold Spring Harbor Laboratory Press, Cold Spring Harbor, New York, pp. 467–494.

Parsons, P.A. (1995). Inherited stress resistance and longevity: a stress theory of ageing. *Heredity* 75: 216–221.

Parsons, P.A. (1996). Rapid development and a long life: an association expected under a stress theory of aging. *Experientia* (Basel). 52: 643–646.

Parsons, P.A. (1997). Stress-resistance genotypes, metabolic efficiency and interpreting evolutionary change. In: *Environmental Stress, Adaptation and Evolution.* (Bijlsma, R., and Loeschcke, V.). Birkhäuser, Basel, Switzerland, pp. 291–305.

Patil, C. and Walker, P. (2001). Intracellular signaling from the endoplasmic reticulum to the nucleus: the unfolded protein response in yeast and mammals. *Curr. Opin. Cell Biol.* 13: 349–355.

Pauli, D. and Tissières, A. (1990). Developmental expression of the heat shock genes in *Drosophila melanogaster*. In: *Stress Proteins in Biology and Medicine* (Morimoto, R.I., Tissières, A., and Georgopoulos, C., Eds.). Cold Spring Harbor Laboratory Press, Cold Spring Harbor, New York, pp. 361–378.

Pauli, D., Arrigo, A.-P., and Tissières, A. (1992). Heat shock response in *Drosophila*. *Experientia* 48: 623–628.

Pelham, H.R.B. (1990). Functions of the hsp70 protein family: An overview. In: *Stress Proteins in Biology and Medicine* (Morimoto, R.I., Tissières, A., and Georgopoulos, C., Eds.). Cold Spring Harbor Laboratory Press, Cold Spring Harbor, New York, pp. 287–299.

Perrimon, N. (1994). Signalling pathways initiated by receptor tyrosine kinase in *Drosophila. Curr. Opin. Cell. Biol.* 6: 620–626.

Pfanner, N. (1990). Mitochondrial protein import: unfolding and refolding of precursor proteins. In: *Stress Proteins: Induction and Function* (Schlesinger, M.J., Santoro, M.G., and Garaci, E., Eds.). Springer-Verlag, Heidelberg, pp. 71–77.

Pinkus, R., Weiner, L.M., and Daniel, V. (1995). Role of quinone-mediated generation of hydroxyl radicals in the induction of glutathione-S-transferase gene expression. *Biochemistry* 34: 81–88.

Plesofsky-Vig, N. (1996). The heat shock proteins and the stress response. *The Mycota 3, Biochemistry and Molecular Biology* (Brambl, R. and Marzlut, G.A., Eds.). Springer-Verlag, Heidelberg, pp. 171–187.

Pletcher, S.D., Macdonald, S.J., Marguerie, R., Certa, U., Stearns, S.C., Goldstein, D.B., and Partridge, L. (2002). Genome-wide transcript profiles in aging and calorically restricted *Drosophila melanogaster. Curr. Biol.* 12: 712–723.

Pohjanvirta, R. and Tuomisto, J. (1994). Short-term toxicity of 2,3,7,8-tetrachlorodibenzo-p-dioxin in laboratory animals: effects, mechanisms, and animal models. *Pharmacol. Rev.* 46: 483–549.

Polyak, K., Xia, Y., Zweier, J.L., Kinzler, K.W., and Vogelstein, B. (1997). A model for p53-induced apoptosis. *Nature* 389: 300–305.

Posthuma, L. (1992). Genetic ecology of metal tolerance in Collembola. Ph.D. thesis, Vrije University, Amsterdam, the Netherlands.

Posthuma, L. and van Straalen, N.M. (1993). Heavy-metal adaptation in terrestrial invertebrates. A review of occurrence, genetics, physiology, and ecological consequences. *Comp. Biochem. Physiol.* 106 C: 11–38.

Puga, A., Nebert, D.W., and Carrier, F. (1992). Dioxin induces expression of c-fos and c-jun proto-oncogenes and a large increase in transcription factor AP-1. *DNA Cell Biol.* 11: 269–281.

Quillet-Mary, A., Jaffrézon, J.P., Mansat, V., Bordier, C., Naval, J., and Laurent, G. (1997). Implication of mitochondrial hydrogen peroxide generation in ceramide induced apoptosis. *J. Biol. Chem.* 272: 21388–21395.

Rabindran, S.K., Wisniewski, J., Li, L., Li, G.C., and Wu, C. (1994). Interaction between heat shock factor and hsp70 is insufficient to suppress induction of DNA-binding activity in vivo. *Mol. Cell. Biol.* 14: 6552–6560.

Ramos-Morales, P. and Rodriguez-Arnaiz, R. (1995). Genotoxicity of two arsenic compounds in germ cells and somatic cells of *Drosophila melanogaster. Environ. Mol. Mutagen.* 25: 288–299.

Ranasinghe, C., Headlam, M., and Hobbs, A.A. (1997). Induction of the mRNA for CYP6B2, a pyrethroid inducible cytochrome P450 in *Helicoverpa armigera* (Hubner) by dietary monoterpenes. *Arch. Insect Biochem. Physiol.* 34: 99–109.

Ranasinghe, C. and Hobbs, A.A. (1999). Induction of cytochrome P450 CYP6B7 and cytochrome b5 mRNAs from *Helicoverpa armigera* (Hubner) by pyrethroid insecticides in organ culture. *Insect Mol. Biol.* 8: 443–447.

Raymond, M., Callaghan, A., Fort, P., and Pasteur, N. (1991). Worldwide migration of amplified insecticide resistance genes in mosquitoes. *Nature* 350: 151–153.

Rice-Evans, C.A. (1994). Formation of free radicals and mechanisms of action in normal biochemical processes and pathological states. In: *Free Radical Damage and Its Control* (Rice-Evans, C.A. and Burdon, R.H., Eds.). Elsevier Science, Amsterdam, the Netherlands, pp. 131–153.

Richter, C. and Schweizer, M. (1997). Oxidative stress in mitochondria. In: *Oxidative Stress and the Molecular Biology of Antioxidant Defenses* (Scandalios, J.G., Ed.). Cold Spring Harbor Laboratory Press, Cold Spring Harbor, New York, pp. 169–200.

Rodriguez, A., Chen, P., Oliver, H., and Abrams, J.M. (2002). Unrestrained caspase-dependent cell death caused by loss of Diap 1 function requires the *Drosophila* Apaf-1 homolog, Dark. *EMBO J.* 21: 2189–2197.

Roesijadi, G. (1996). Metallothionein and its role in toxic metal regulation. *Comp. Biochem. Physiol.* 113 C: 117–123.

Roméo, M., Bennani, N., Gnassia-Barelli, M., Lafaurie, M., and Girard, J.P. (2000). Cadmium and copper display different responses towards oxidative stress in the kidney of the sea bass *Dicentrarchus labrax. Aquat. Toxicol.* 48: 185–194.

Rothman, J.E. (1989). Polypeptide chain binding proteins: catalysts of protein folding and related processes in cells. *Cell* 59: 591–601.

Rouse, J., Cohen, P., Trigon, S., Morange, M., Alonso-Llamazares, A., Zamarillo, D., Hunt, T., and Nebreda, A.R. (1994). A novel kinase cascade triggered by stress and heat shock that stimulates MAPKAP kinase-2 and phosphorylation of the small heat shock proteins. *Cell* 78: 1027–1037.

Rubin, D.M., Mehta, A.D., Zhu, J., Shohan, S., Chen, X., Wells, Q.R., and Palter, K.B. (1993). Genomic structure and sequence analysis of *Drosophila melanogaster* HSC70 genes. *Gene* 128: 155–163.

Rutherford, S.L. and Zuker, C.S. (1994). Protein folding and regulation of signaling pathways. *Cell* 79: 1129–1132.

Ryan, K.M., Phillips, A.C., and Vousden, K.H. (2001). Regulation and function of the p53 tumor suppressor protein. *Curr. Opin. Cell Biol.* 13: 332–337.

Salvesen, G.S. and Dixit, V.M. (1997). Caspases: intracellular signaling by proteolysis. *Cell* 91: 443–446.

Sanders, B.M. (1990). Stress proteins: potential as multitiered biomarkers. In: *Environmental Biomarkers* (McCarthy, J.F. and Shugart, R.L., Eds.). Lewis Publishers, Chelsea, MI, pp. 165–191.

Sarge, K.G., Murphy, S.P., and Morimoto, R.I. (1993). Activation of heat shock gene transcription by HSF1 involves oligomerisation, acquisition of DNA binding activity, and nuclear localisation and can occur in the absence of stress. *Mol. Cell. Biol.* 13: 1392–1407.

Sato, M. and Bremner, I. (1993). Oxygen free radicals and metallothionein. *Free Rad. Biol. Med.* 14: 325–337.

Schlesinger, M.J. (1990). The ubiquitin system and the heat shock response. In: *Stress Proteins: Induction and Function* (Schlesinger, M.J., Santoro, M.G., and Garaci, E., Eds.). Springer-Verlag, Heidelberg, pp. 81–88.

Schmidt, K.N., Amstad, P., Cerutti, P., and Baeuerle, P.A. (1995). The roles of hydrogen peroxide and superoxide as messengers in the activation of transcription factor NF-κB. *Chem. Biol.* 2: 13–22.

Schwartz, M.A., Lazo, J.S., Yalowich, J.C., Reynolds, I., Kagan, V.E., Tyurin, V., Kim, Y.-M., Watkins, S.C., and Pitt, B.R. (1994). Cytoplasmic metallothionein over-expression protects NIH 3T3 cells from tert-butyl hydroperoxide toxicity. *J. Biol. Chem.* 269: 15238–15243.

Scott, J.G., Liu, N., and Wen, Z. (1998). Insect cytochromes P450: diversity, insecticide-resistance, and tolerance to plant toxins. *Comp. Biochem. Physiol. C: Comp. Pharmacol. Toxicol.* 121: 147–155.

Sewell, A.K., Yokoya, F., Yu, W., Miyagawa, T., Murayama, T., and Winge, D.R. (1995). Mutated yeast heat shock transcription factor exhibits elevated basal transcriptional activation and confers metal resistance. *J. Biol. Chem.* 270: 25079–25086.

Sheng, M.E., Thompson, M.A., and Greenberg, M.E. (1991). CREB: a Ca^{2+}-regulated transcription factor phosphorylated by calmodulin-dependent kinases. *Science* 252: 1427–1430.

Shi, L., Sawada, M., Sester, U., and Carlson, J.C. (1994). Alterations in free radical activity in aging *Drosophila. Exp. Gerontol.* 29: 575–584.

Shi, Y., Mosser, D.D., and Morimoto, R.I. (1998). Molecular chaperones as HSF-1-specific transcriptional repressors. *Genes Dev.* 12: 654–666.

Shigenaga, M.K., Hagen, T.M., and Ames, B.N. (1994). Oxidative damage and mitochondrial decay in aging. *Proc. Natl. Acad. Sci. USA* 91: 10771–10778.

Shirley, M.D.F. and Silbly, R.M. (1999). The genetic basis of a between environment trade-off involving resistance to cadmium in *Drosophila melanogaster. Evolution* 53: 826–836.

Silar, P., Butler, G., and Thiele, D.J. (1991). Heat shock transcription factor activates transcription of the yeast metallothionein gene. *Mol. Cell. Biol.* 11: 1232–1238.

Sohal, R.S., Agarwal, A., Agarwal, S., and Orr, W.C. (1995). Simultaneous overexpression of copper- and zinc-containing superoxide dismutase and catalase retards age-related oxidative damage and increases metabolic potential in *Drosophila melanogaster. J. Biol. Chem.* 270: 15671–15674.

Sohal, R.S. (2002). Role of oxidative stress and protein oxidation in the aging process. *Free Rad. Biol. Med.* 33: 37–44.

Sohal, R.S., Mockett, R.J., and Orr, W.C. (2002). Mechanisms of aging: an appraisal of the oxidative stress hypothesis. *Free Rad. Biol. Med.* 33: 575–586.

Sok, J., Calfon, M., Lu, J., Lichtlen, P., Clark, S.G., and Ron, D. (2001). Arsenite-inducible RNA-associated protein (A1RAP) protects cells from arsenite toxicity. *Cell Stress Chaperones* 6: 6–15.

Sörensen, J.G. and Loeschcke, V. (2002). Decreased heat-shock resistance and down-regulation of Hsp70 expression with increasing age in adult *Drosophila melanogaster. Funct. Ecol.* 16: 379–384.

Sorger, P.K. (1991). Heat shock factor and the heat shock response. *Cell* 65: 363–366.

Spiegelman, V.S., Fuchs, S.Y., and Belitsky, G.A. (1997). The expression of insecticide resistance-related cytochrome P450 forms is regulated by molting hormone in *Drosophila melanogaster. Biochem. Biophys. Res. Commun.* 232: 304–307.

Stadtman, E.R. (1992). Protein oxidation and aging. *Science* 257: 1220–1224.

Stege, G.J.J., Li, L., Kampinga, H.H., Konings, A.W.T., and Li, G.C. (1994). Importance of the ATP-binding domain and nucleolar localization domain of hsp72 in the protection of nuclear proteins against heat-induced aggregation. *Exp. Cell. Res.* 214: 279–284.

Stepanova, L., Leng, X., Parker, S.B., and Harper, J.W. (1996). Mammalian p50 Cdc37 is a protein kinase targeting subunit of hsp90 that binds and stabilizes CdK4. *Genes Dev.* 10: 1491–1502.

Stephenson, G.F., Chan, H.M., and Cherian, M.G. (1994). Copper-metallothionein from the toxic milk mutant mouse enhances lipid peroxidation initiated by an organic hydroperoxide. *Toxicol. Appl. Pharmacol.* 125: 90–96.

Storz, G. and Polla, B.S. (1996). Transcriptional regulation of oxidative stress-inducible genes in prokaryotes and eukaryotes. In: *Stress-Inducible Cellular Responses* (Feige, U., Morimoto, R.I., Yahara, I., and Polla, B.S., Eds.). Birkhäuser, Basel, Switzerland, pp. 239–254.

Sugiyama, M. (1994). Role of cellular antioxidants in metal-induced damage. *Cell. Biol. Toxicol.* 10: 1–22.

Tanguay, R.M., Joanisse, D.R., Inaguma, Y., and Michaux, S. (1999). Small heat shock proteins: in search of functions in vivo. In: *Environmental Stress and Gene Regulation* (Storey, K.B., Ed.). Bios Scientific, Oxford, pp. 125–138.

Tatar, M., Kopelman, A., Epstein, D., Tu, M.-P., Yin, C.-M., and Garofalo, R.S. (2001). A mutant *Drosophila* insulin receptor homolog that extends life-span and impairs neuroendocrine function. *Science* 292: 107–110.

Thiele, D.J. (1992). Metal regulated transcription in eukaryotes. *Nucl. Acids Res.* 20: 1183–1191.

Timbrell, J.A. (1991). *Principles of Biochemical Toxicology* (2nd ed.). Taylor & Francis, London.

Tomita, T., Liu, N., Smith, F.F., Sridhar, P., and Scott, J.G. (1995). Molecular mechanisms involved in increased expression of cytochrome P450 responsible for pyrethroid resistance in the house fly *Musca domestica*. *Insect Mol. Biol.* 4: 135–140.

Tong, W.-M., Galendo, D., and Wang, Z.-Q. (2000). Role of DNA break-sensing molecule poly (ADP-ribose) polymerase (PARP) in cellular function and radiation toxicity. In: *Cold Spring Harbor Symposium on Quantitative Biology: Biological Responses to DNA Damage*, Vol. 65. Cold Spring Harbor Laboratory Press, Cold Spring Harbor, New York, pp. 583–589.

Trump, B.F. and Berezesky, I.K. (1995). Calcium-mediated cell injury and cell death. *FASEB J.* 9: 219–228.

Tsutsayeva, A.A. and Sevryukova, L.G. (2001). Effect of cold exposure on survival and stress protein expression of *Drosophila melanogaster* at different development stages. *CryoLetters* 22: 145–150.

Valls, M., Bofill, R., Romero-Isart, N., Gonzalez-Duarte, R., Abian, J., Carrascal, P., Gonzales-Duarte, M., Capdevila, M., and Atrian, S. (2000). The *Drosophila* MTN: a metazoan copper-thionein related to fungal forms. *FEBS Lett.* 467: 189–194.

van der Oost, R. (1998). Fish biomarkers for inland water pollution. Ph.D. thesis, Vrije Universiteit, Amsterdam, the Netherlands.

van Straalen, N.M. and Verkleij, J.A.C. (1991). *Leerboek Oecotoxicologie*. VU Uitgeverij, Amsterdam, the Netherlands.

van Straalen, N.M. (2003). Ecotoxicology becomes stress ecology. *Environ. Sci. Technol.* 37: 324A–330A.

Voellmy, R. (1996). Sensing stress and responding to stress. In: *Stress-Inducible Cellular Responses* (Feige, U., Morimoto, R.I., Yahara, I., Polla, B.S., Eds.). Birkhäuser, Basel, Switzerland, pp. 121–137.

Wang, Z. and Templeton, D.M. (1998). Induction of c-fos proto-oncogene in mesangial cells by cadmium. *J. Biol. Chem.* 273: 73–79.

Watowich, S.S. and Morimoto, R.I. (1988). Complex regulation of heat shock and glucose responsive genes in human cells. *Mol. Cell. Biol.* 8: 393–405.

Waxman, D.J. (1988). Interactions of hepatic cytochromes P-450 with steroid hormones. *Biochem. Pharmacol.* 37: 71–84.

Wei, J. and Hendershot, L.M. (1996). Protein folding and assembly in the endoplasmic reticulum. In: *Stress-Inducible Cellular Responses* (Feige, U., Morimoto, R.I., Yahara, I., Polla, B.S., Eds.). Birkhäuser, Basel, Switzerland, pp. 41–55.

Welch, W.J. and Mizzen, L.A. (1988). Characterization of the thermotolerant cell. II. Effects on the intracellular distribution of heat shock protein 70, intermediate filaments, and small ribonucleoprotein complexes. *J. Cell. Biol.* 106: 1117–1130.

Welch, W.J. (1990). The mammalian stress response: cell physiology and biochemistry of stress proteins. In: *Stress Proteins in Biology and Medicine* (Morimoto, R.I., Tissières, A., and Georgopoulos, C., Eds.). Cold Spring Harbor Laboratory Press, Cold Spring Harbor, New York, pp. 223–278.

Welch, W.J. (1992). Mammalian stress response: cell physiology, structure/function of stress proteins, and implications for medicine and disease. *Physiol. Rev.* 72: 1063–1081.

Welch, W.J. (1993). How cells respond to stress. *Sci. Am.* May 1993: 34–41.

Wertz, I.E., and Hanley, M.R. (1996). Diverse molecular provocation of programmed cell death. *Trends Biochem. Sci.* 21: 359–364.

Whitlock J.P., Jr. and Denison, M.S. (1995). Induction of cytochrome P450 enzymes that metabolize xenobiotics. In: *Cytochrome P450: Structure, Mechanism, and Biochemistry* (2nd ed.) (Ortiz de Montellano, P.R., Ed.). Plenum Press, New York, pp. 367–390.

Wickens, A.P. (2001). Ageing and the free radical theory. *Respir. Physiol.* 128: 379–391.

Winegarden, N.A., Wong, K.S., Sopta, M., and Westwood, J.T. (1996). Sodium salicylate decreases intracellular ATP, induces both heat shock factor binding and chromosomal puffing, but does not induce hsp70 gene transcription in *Drosophila. J. Biol. Chem.* 271: 26971–26980.

Winyard, P.G., Morris, C.J., Winrow, V.R., Zaidi, M., and Blake, D.R. (1994). Free radical pathways in the inflammatory response. In: *Free Radical Damage and Its Control* (Rice-Evans, C.A. and Burdon, R.H., Eds.). Elsevier Science, Amsterdam, the Netherlands, pp. 361–383.

Wolin, M.S. and Mohazzab-H, K.M. (1997). Mediation of signal transduction by oxidants. In: *Oxidative Stress and the Molecular Biology of Antioxidant Defenses* (Scandalios, J.G., Ed.). Cold Spring Harbor Laboratory Press, Cold Spring Harbor, New York, pp. 21–48.

Wu, C., Zimarino, V., Taal, C., Walker, B., and Wilson, S. (1990). Transcriptional regulation of heat shock genes. In: *Stress Proteins in Biology and Medicine* (Morimoto, R.I., Tissières, A., and Georgopoulos, C., Eds.). Cold Spring Harbor Laboratory Press, Cold Spring Harbor, New York, pp. 429–442.

Wu, C., Clos, J., Giorgi, G., Haroun, R.I., Kim, S.-J., Rabindran, S.K., Westwood, J.T., Wisniewski, J., and Yim, G. (1994). Structure and regulation of heat shock transcription factor. In: *The Biology of Heat Shock Proteins and Molecular Chaperones* (Morimoto, R.I., Tissières, A., and Georgopoulos, C., Eds.). Cold Spring Harbor Laboratory Press, Cold Spring Harbor, New York, pp. 395–416.

Wu, C. (1995). Heat shock transcription factors: structure and regulation. *Annu. Rev. Cell Dev. Biol.* 11: 441–469.

Xia, W. and Voellmy, R. (1997). Hyperphosphorylation of heat shock transcription factor-1 is correlated with transcriptional competence and slow dissociation of active factor trimers. *J. Biol. Chem.* 272: 4094–4102.

Yost, H.J., Petersen, R.B., and Lindquist, S. (1990). Posttranscriptional regulation of heat shock protein synthesis in *Drosophila*. In: *Stress Proteins in Biology and Medicine* (Morimoto, R.I., Tissières, A., and Georgopoulos, C., Eds.). Cold Spring Harbor Laboratory Press, Cold Spring Harbor, New York, pp. 379–409.

Yu, B.P. (1994). Cellular defenses against damage from reactive oxygen species. *Physiol. Rev.* 74: 139–162.

Yu, C.W., Chen, J.H., and Lin, L.Y. (1997). Metal-induced metallothionein gene expression can be inactivated by protein kinase C inhibitor. *FEBS Lett.* 420: 69–73.

Yu, S.P., Canzoniero, L.M.T., and Choi, D.W. (2001). Ion homeostasis and apoptosis. *Curr. Opin. Cell Biol.* 13: 405–411.

Zaccolo, M., Magelhães, P., and Pozzan, T. (2002). Compartmentalisation of cAMP and Ca^{2+} signals. *Curr. Opin. Cell Biol.* 14: 160–166.

Zachariassen, K.E. (1991). The water relations of overwintering insects. In: *Insects at Low Temperature* (Lee, R.E., Jr. and Denlinger, D.L., Eds.). Chapman & Hall, New York, pp. 47–63.

Zhang, B., Egli, D., Georgiev, O., and Schaffner, W. (2001). The *Drosophila* homolog of mammalian zinc finger factor MTF-1 activates transcription in response to heavy metals. *Mol. Cell. Biol.* 21: 4505–4514.

Zhang, J. and Xu, M. (2002). Apoptotic DNA fragmentation and tissue homeostasis. *Trends Cell Biol.* 12: 84–89.

Zhao, R., Gish, K., Murphy, M., Yin, Y., Notterman, D., Hoffman, W.H., Tom, E., Mack, D.H., and Levine, A.J. (2000). The transcriptional program following p53 activation. In: *Cold Spring Harbor Symposium on Quantitative Biology: Biological Responses to DNA Damage*, Vol. 65. Cold Spring Harbor Laboratory Press, Cold Spring Harbor, New York, pp. 483–488.

Zhong, M., Orosz, A., and Wu, C. (1998). Direct sensing of heat and oxidation by *Drosophila* heat shock transcription factor. *Mol. Cell* 2: 101–108.

Zimmermann, K.C., Ricci, J.-E., Droin, N.M., and Green, D.R. (2002). The role of ARK in stress-induced apoptosis in *Drosophila* cells. *J. Cell. Biol.* 156: 1077–1087.

Index

A